APPLICATION OF ELECTRO-ULTRAFILTRATION (EUF) IN AGRICULTURAL PRODUCTION

Proceedings of the First International Symposium on
the Application of Electro-Ultrafiltration in Agricultural Production,
organized by the Hungarian Ministry of Agriculture and
the Central Research Institute for Chemistry of the Hungarian
Academy of Sciences, Budapest, May 6-10, 1980

W0193124

edited by

K. NÉMETH

Reprinted from *Plant and Soil*, vol. 64, no. 1 (1982)

1982

MARTINUS NIJHOFF / DR W. JUNK PUBLISHERS
THE HAGUE / BOSTON / LONDON

Distributors:

for the United States and Canada

Kluwer Boston, Inc.
190 Old Derby Street
Hingham, MA 02043
USA

for all other countries

Kluwer Academic Publishers Group
Distribution Center
P.O. Box 322
3300 AH Dordrecht
The Netherlands

Library of Congress Cataloging in Publication Data CIP

ISBN-13: 978-90-247-2641-7   e-IBN-13: 978-94-009-7559-0
DOI: 10.1007/978-94-009-7559-0

Application of electro-ultrafiltration (EUF) in agricultural production

Contents

Application of electro-ultrafiltration (EUF) in agricultural production

Preface

The First International Symposium on the Application of Electro-ultrafiltration in Agricultural Production was held in Budapest (Hungary) from May 6–10, 1980. Over 100 individuals from 15 different countries in Europe, Asia and Africa took part and 35 papers on experience with this method were presented. What was the idea behind this Symposium?

The procedure of electro-ultrafiltration as suggested by Bechold in 1925 has been further developed and improved over the past ten years. The introduction of varying voltage and temperature during the extraction process gives this method a considerable advantage in plant nutrient analysis, as it is now possible to obtain several nutrient fractions of different plant availability in one extraction run and to determine at the same time other soil properties such as K selectivity of clay minerals, content of K selective clay minerals, content of $CaCO_3$ etc. These data, in fact, make it possible to characterize a soil comprehensively and thus to ensure optimal plant nutrition, so meeting the economic requirements of agricultural production.

In 1974 and on the initiative of Dr. Wiklicky the Tulln Sugar Factory in Austria decided to introduce the EUF procedure in its soil testing laboratory for routine analysis. This meant a considerable step forward in the development of EUF. About 15 to 20 thousand soil samples every year are analysed at Tulln. The practical experience gained made it possible to improve the EUF procedure. In 1974 the UEF method was also introduced in Hungary by Dr. Eifert; he and his collaborators obtained remarkable results in grape production.

With the assistance of Dr. Wiklicky and his co-workers, a modern EUF soil testing laboratory has been established in Ormož, Yugoslavia, with the aim of improving fertilizer use on sugar beet. The EUF procedure is now being used in over 20 countries worldwide either on test or for the practical improvement of fertilizer advice. The worldwide interest in the method suggested that the time was ripe to collate existing information in a Symposium. The Hungarian Ministry of Agriculture favoured the proposal and preparations for the Symposium to be held in the Hungarian capital commenced in 1979.

The scientific content of the EUF Symposium was prepared at the Büntehof Agricultural Research Station in Hannover. Grateful acknowledgements are extended to Prof. Beringer, the Director of the Büntehof, and his predecessor Prof. Mengel, who contributed so generously to the development of the EUF programme. I am also very much indebted to the staff members of the Büntehof and in particular to the translator Mrs. Labrenz for the unfailing help they have always given me.

The Symposium was organized by the Hungarian Ministry of Agriculture and the Central Research Institute for Chemistry of the Hungarian Academy of Sciences. This arduous task was coordinated by Dr. Varju to whom I wish to express my appreciation for his whole-hearted assistance. After the Symposium the participants had the opportunity to visit the vineyards and cellars of the 'Balatonboglár' state holding which proved of great interest. The First International EUF Symposium owes its success to the excellent cooperation between the Büntehof Agricultural Research Station and the Hungarian Ministry of Agriculture, including the Central Research Institute for Chemistry of the Hungarian Academy of Sciences.

The objective of the Symposium was to present the first results and experience gained with the EUF procedure, so that the papers delivered are only a survey of the present-day state of knowledge. The bulk of the work of editing, including the selection of the articles, fell on Prof. Dr. van Diest, Wageningen, for whose valuable advice and untiring efforts I wish to record my special thanks. Some of the papers presented at the Symposium have already been published elsewhere and they will not appear in this issue.

Interest in the EUF procedure continues to grow. In Germany, for instance, Südzucker AG recently decided to organize a modern EUF soil-testing laboratory with the assistance of the Tulln Sugar Factory. The worldwide EUF activities and the positive results of the First International Symposium on EUF are an encouragement and a challenge to present an up-to-date account in the near future.

October, 1981
Hannover (Fed. Rep. Germany)   K. NÉMETH

Electro-ultrafiltration of aqueous soil suspension with simultaneously varying temperature and voltage

K. NÉMETH

Büntehof Agricultural Research Station Hannover, Federal Republic of Germany

Key words Fertilizer requirement Nutrient availability Nutrient fractions Nutrient reserves
Soil analysis Soil solution

Summary The procedure of electro-ultrafiltration (EUF) with simultaneously varying voltage and temperature for the determination of nutrient fractions in soil as well as for the characterization of important soil properties such as kind of clay, content of K-selective clay minerals, of $CaCO_3$ etc. is described. The interpretation of the EUF–N, EUF-P and EUF-K fractions is discussed in detail with regard to their availabilities. Moreover the influence of soil properties and soil management on the EUF nutrient fractions is discussed.

For characterization of the EUF values needed for optimum plant nutrition, it is suggested to combine the EUF values at 20°C and 200 V on the one hand (easily available fraction) and the EUF values at 80°C and 400 V on the other (nutrient reserves). The higher, in fact, the nutrient contents at 80°C and 400 V are the lower can be the EUF values at 20°C and 200 V. In contrast to conventional methods of soil analysis, the nutrient contents determined by EUF are expressed in mg/100 g soil/time unit of desorption (min) at given temperature and voltage.

The EUF procedure proved highly advantageous for an assessment of the fertilizer requirements of soils as this method allows for the determination of soil properties such as kind of clay, content of K-selective clay minerals, of $CaCO_3$ and of easily desorbable heavy metals in one single operation. These soil properties play a decisive role in the sorption of K and P.

Introduction

In agricultural practice, soil analysis is expected to answer the following questions:

a) What is the amount of nutrients in the soil which can be supplied to the plant root in the course of a vegetation period?

b) Which of the changes in these easily available amounts are due to crop withdrawal, leaching, weathering *etc.*?

c) Which are the amounts of nutrients to be applied to the soil in order to raise the easily available amounts to the required values (fertilizer requirements)?

It will be explained how this information can be obtained through the use of electro-ultrafiltration (EUF) when voltage and temperature are varied during the extraction process.

Procedure of extraction

Several other publications[4,5,6] give a full description of the component processes of electro-ultrafiltration (dialysis, ultrafiltration and electrodialysis) and the development of electro-ultrafiltration since 1925[1], so that these details need not be discussed further here. The present paper will be concerned with the consequence of introducing variations in voltage and temperature during the extraction process which has resulted into a considerable increase in the usefullness of the EUF procedure.

Significance of simultaneously altering temperature and voltage

The speed of ion migration in electro-ultrafiltration is proportional to the field strength and inversely proportional to the frictional forces. Furthermore, at a given field strength ion migration increases with increasing concentration. Temperature also plays a decisive role in EUF extraction. When the temperature is raised, higher amounts of nutrients are extracted at constant voltage within a unit of time[5,6].

It is, however, necessary to examine which additional nutrient fractions are withdrawn from a soil when temperature and voltage are increased and whether or not these fractions are important from a standpoint of plant nutrition.

For the determination of the fractions of nutrients which can most easily be made available to plant roots in the course of a vegetation period it is suggested to apply a 30 min extraction at 200 V and 20°C. The nutrient reserves will be assessed when the temperature is raised from 20°C to 80°C and the extraction continued for another 5 min or more. Extending the extraction for 5 min at 80°C and 400 V proved most suitable for practical purposes.

The significance of altering voltage and temperature as well as working at constant voltage and temperature is demonstrated in Fig. 1 for the extraction of the nutrient K. This figure shows the EUF–K curves of a humid tropical soil with kaolinitic clay minerals (exchangeable K = 13.6 mg/100 g soil) and of a grey-brown luvisol with illitic clay minerals (exchangeable K = 16.9 mg/100 g soil). The amounts of exchangeable K in the kaolinitic soil are quantitatively desorbed by means of EUF at 200 V and 20°C within 20 min ($t_{0.5}$ = 4 min). No more K ions are released when voltage and temperature are raised. It may, therefore, be concluded that this soil has no K reserves (Fig. 1, on the left). In the illitic soil, on the contrary, only part of the exchangeable K ions, namely 8.6 from 16.9 mg, are released at 200 V and 20°C within 30 min. The maximally desorbable K amounts (dm) of 17.9 mg at 200 V and 20°C are almost equivalent to the contents of exchangeable K, the half time value being 31 min. When voltage and temperature are raised from 200 V and 20°C to 400 V and 80°C, the illitic soil, unlike the kaolinitic soil, continues to release K ions which attests to the K reserves of this soil. In this soil, therefore, the sum of the maximally desorbable K amounts at 200 V and 20°C and at 400 V and 80°C is higher (28.9 mg) than the contents of exchangeable K (16.9 mg).

K desorption at 400 V and 80°C, in contrast to the results obtained when varying voltage and temperature are applied, is similar in both soils (Fig. 1, on the right). Only the half time values of 1.4 (for the kaolinitic soil) and 7.2 (for the illitic soil) indicate different characteristics of desorption. When voltage and temperature are varied, the EUF–K curve for the kaolinitic soils shows a completely different course than for the illitic soils. The maximally desorbable K amounts of illitic soils at 400 V and 80°C are lower (21.6 mg) than the sum of maximally desorbable K amounts when voltage and temperature are varied (28.9 mg). For the assessment of the K reserves it is therefore more suitable to raise temperature and voltage during the extraction than to work at constant temperature and voltage.

In which way high K reserves are indicated by varying voltage and temperature is also demonstrated in Fig. 2. This figure shows the EUF–K curves of an alluvial soil (illitic clay minerals) with a high content of exchangeable K (79 mg/100 g). Despite the high contents of exchangeable K, only small amounts of K were desorbed at 20°C and 200 V. K desorption is, however, considerably accelerated, when voltage and temperature are raised, thus indicating a high K reserve. An increase in voltage alone (20°C and 400 V) is not sufficient to indicate this high reserve.

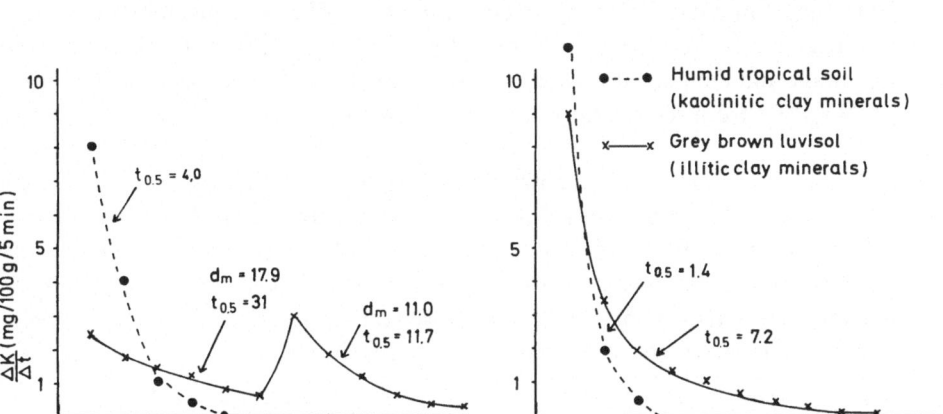

Fig. 1. EUF–K desorption rates at constant temperature (80°C) and constant voltage (400), on the one hand, and at varying voltage (200, 400 V) and temperature (20°, 80°) on the other hand.

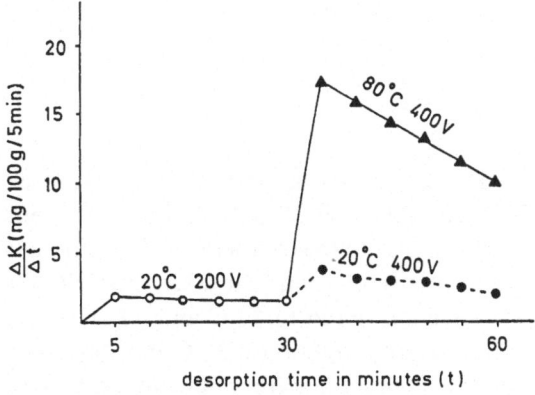

Fig. 2. EUF–K desorption rates of an alluvial soil (60% clay; illitic clay minerals) with a high content of exchangeable K (79 mg/100 g).

Interpretation of EUF results for plant nutrition

EUF-nitrogen fractions

$EUF-NH_4$ fractions The EUF–NH_4 contents of well-drained soils are lower than the EUF–NO_3 fractions. In soils well supplied with K the EUF–NH_4 contents are often so low that they are hardly measurable. In rice soils, however, the EUF–NH_4 contents can be rather substantial and they are correlated with the N absorbed by the rice plants[11].

In well-drained soils low in K, the EUF–NH_4 contents increase with increasing contents of K-selective clay minerals[6]. The desorption rate of NH_4^+ ions becomes higher with increasing voltage and temperature. Thus, an increase in voltage and temperature seems advisable to obtain the necessary information on N release from the exchangeable NH_4 reserves.

EUF–NO_3 fractions It is known from earlier investigations that within 15–20 min the NO_3^- ions in a solution are quantitatively transferred by electro-ultrafiltration at 20°C and 200 V. An experiment was carried out to study the influence of the presence of a soil on the transfer of NO_3^- ions. A grey-brown luvisol was treated with KNO_3 and subsequently extracted by EUF for NO_3 analysis in the extracts. The results are given in Table 1.

Table 1. EUF–NO_3 contents of a grey-brown luvisol before and immediately after applying KNO_3 (0.5 mg N/100 g soil)

Soil status	EUF–NO_3 (mg N/100 g)	
	20°C, 200 V (0–30 min)	80°C, 400 V (30–60 min)
Before KNO_3 application	1.44	0.80
After KNO_3 application	1.95	0.82

It can be seen that the NO_3 applied to the soil can be quantitatively recovered in the EUF fractions obtained at 20°C and 200 V. The transfer of the NO_3^- ions from a soil suspension is, therefore, very similar to the transfer from a solution. It is not easy to provide an explanation for the finding that additional NO_3 is extracted, when temperature and voltage are raised to 80°C and 400 V for an additional 30 min period after an initial 30 min extraction at 20°C and 200 V. It is conceivable that low-molecular organic N compounds are anodically oxidized to NO_3, but there is still no actual evidence for this assumption.

The EUF–NO_3 fraction at 80°C and 400 V was not influenced by KNO_3 application. It was, therefore, investigated which factors influence the EUF–NO_3 fractions. For this purpose, a field experiment (Büntehof) was carried out with different soils (1-m^2 plots) which for a 10-year period had been managed alike (uniform N-application, uniform crop rotations). Table 2 shows the EUF–NO_3 fractions as well as the Kjeldahl-N values which were obtained in the above experiment for the various soil types included in the trial. It can be seen that the EUF–NO_3 values obtained after uniform N applications over a 10-year period show soil-specific differences. The data furthermore show that a correlation exists between Kjeldahl-N values and EUF–NO_3 contents.

Table 2. Influence of variation in soil type on the $EUF-NO_3$ value after uniform N applications over a 10-year period

Soil type	$EUF-NO_3$ (mg N/100 g)		Kjeldahl-N (mg/100 g)
	20°C, 200 V (0–30 min)	80°C, 400 V (30–60 min)	
Chernozem	1.10	0.80	142
Pararendzina	0.80	0.51	50
Grey-brown luvisol	1.00	0.54	91
Alluvial soil	0.80	0.60	116
Humic sand	1.50	2.06	290

$EUF-NO_3$ at 20°C = 0.0029 Kjeldahl-N + 0.63

$r = 0.93**$

$EUF-NO_3$ at 80°C = 0.007 Kjeldahl-N − 0.057

$r = 0.97***$

These correlations between $EUF-NO_3$ and Kjeldahl-N cannot be found very often, as they require uniform climatic and management conditions. A certain degree of correlation was only observed between the $EUF-NO_3$ fractions at 80°C and 400 V and Kjeldahl-N[3]. Differences in K and P supply of a soil will already lead to different NO_3 values under uniform cropping (vine) as demonstrated in Table 3. The results indicate that the $EUF-NO_3$ contents decrease with improved K and P supply, as the vine optimally supplied with K and P takes up more NO_3. The $EUF-NO_3$ values at 80°C remain, however, relatively unchanged and thus prove to be more a function of the type of soil involved (except soil No. 228). It is frequently observed that $EUF-NO_3$ values at 80°C increase with improved nutrient supply as can also be seen in Table 3 (soil No. 228).

Application of organic N can also affect the $EUF-NO_3$ values considerably (Table 4). In this table it is, however, clearly demonstrated that application of organic N in the form of slurry raised only the $EUF-NO_3$ values at 20°C.

EUF-extractable N fractions without EUF-NO₃ The cathode and anode filtrates contained more N than the sum of $EUF-NH_4$ and $EUF-NO_3$. This means that other N-containing compounds (serine, glycine, alanine, asparagine, glutamic acid *etc.*) also migrated in the electric field[6].

Table 5 gives information on the influence of soil properties on the EUF-extractable N fractions. It is clearly shown that the content of EUF-extractable N depends on soil properties. Like for the $EUF-NO_3$ values, a correlation appeared to exist between the Kjeldahl-N values and the EUF-extractable N values.

Table 3. EUF–NO$_3$ values as affected by variations in K and P status of soils in field experiments with vine*

Soils No.	EUF–K (20°C)	EUF–P (20°C)	EUF–NO$_3$ (20°C)	EUF–NO$_3$ (80°C)	Kjeldahl-N
	mg/100 g		mg N/100 g		
201	6.0	0.24	2.14	0.70	136.7
200	6.8	0.60	1.90	0.80	142.8
211	10.7	0.56	1.50	0.76	144.2
228	16.0	2.10	1.40	1.10	146.2

* The soil samples were provided by courtesy of Dr. Eifert, Research Institute for Viticulture and Enology, Kecskemét (Hungary).

Table 4. EUF–NO$_3$ values determined after harvest of a wheat crop that had been supplied with organic N in the form of slurry

Soils	EUF–NO$_3$ (mg N/100 g)		Kjeldahl-N (mg/100 g)
	20°C (0–30 min)	80°C (30–60 min)	
Grey-brown luvisol	4.74	0.90	139.4
Chernozem-grey-brown luvisol	6.64	0.92	175.5
Chernozem-grey-brown luvisol (without slurry)	1.06	0.98	163.0

EUF-extractable N at 20°C = 0.009 Kjeldahl-N + 1.12
$r = 0.93**$
EUF-extractable N at 80°C = 0.018 Kjeldahl-N + 0.14
$r = 0.97***$

It should, however, be noted that the soils concerned had received uniform N dressings over a 10-year period at uniform climatic conditions. In contrast to this, Németh[6] and Harrach et al.[3] did not find any correlation between the total N (Kjeldahl-N) and EUF-extractable N in different soils under different management conditions.

Relationship between EUF–NO$_3$ and EUF-extractable values (without EUF–NO$_3$)   At uniform N applications and uniform climatic conditions a correlation

Table 5. EUF-extractable N fractions in different soils that received uniform N-applications over a ten-year period

Soil type	EUF-extractable N (without NO$_3$) (mg N/100 g)		Kjeldahl-N (mg N/100 g)
	20°C, 200 V (0–30 min)	80°C, 400 V (30–60 min)	
Chernozem	2.4	2.2	142
Pararendzina	1.7	1.5	50
Grey-brown luvisol	2.4	2.0	91
Alluvial soil	1.7	2.0	116
Humic sand	4.0	5.8	290

can be observed between the EUF–NO$_3$ values and the EUF-extractable N values. These data were presented in Tables 2 and 5.

EUF–NO$_3$ at 20°C = 0.30 EUF-extractable N at 20°C + 0.29
$r = 0.98$***
EUF–NO$_3$ at 80°C = 0.37 EUF-extractable N at 80°C − 0.10
$r = 0.99$***

Such correlations cannot be found when management conditions differ (*e.g.* when large quantities of slurry are applied)[3].

The application of large quantities of slurry disturbed the equilibrium between the EUF–NO$_3$ values and the values on EUF-extractable N. A balance is not restored, until after the quantities of EUF–NO$_3$ has been incorporated into the biomass or removed by leaching.

Calculation of the easily available quantities of N in a soil　After numerous field experiments conducted on several sites[8], it was found that 1 mg N/100 g soil (measured as EUF–NO$_3$ at 20°C in soils after harvest of cereal crop) is equivalent to approximately 90 kg N/ha. This figure proved highly reliable over the years 1976–1979 in Lower Austria and Burgenland (province of East Austria). Investigations in other European countries and in the USA have, however, revealed that, depending on climatic conditions and soil management, 1 mg EUF–NO$_3$ at 20°C is equivalent to sometimes more and sometimes less than 90 kg fertilizer N/ha.

Further investigations (*cf* Wiklicky[12]) have shown that the easily available quantities of N can more reliably be calculated on the basis of EUF-extractable N (Σ EUF–N) values. In this case, however, 1 mg EUF-extractable total N/100 g soil corresponds to 30 kg fertilizer N/ha. An example is given in Table 6.

Table 6. Calculation of the quantities of easily available N in different soils at uniform N applications

Soils	EUF–NO$_3$ 20°C (mg N/100 g)		Easily available N (kg N/ha)	EUF–N 20°C (mg N/100 g)		Easily available N (kg N/ha)
Chernozem	1.1 × 90	=	99	3.5 × 30	=	105
Pararendzina	0.8 × 90	=	72	2.5 × 30	=	75
Grey-brown luvisol	1.0 × 90	=	90	3.4 × 30	=	102
Alluvial soil	0.8 × 90	=	72	2.5 × 30	=	75
Humic sand	1.50 × 90	=	135	5.5 × 30	=	165

As demonstrated in Table 6 it is possible to calculate the easily available N quantities in soils under uniform N fertilization by using either the EUF–NO$_3$ fractions or the EUF-extractable N fractions. This finding is based on the observation that in this case a close correlation exists between the EUF–NO$_3$ and EUF-extractable N values. Such a correlation depends on an equilibrium existing between the different EUF–N fractions.

This equilibrium can for instance be disturbed by leaching of NO$_3$ or when high amounts of NO$_3$ are added in the form of slurry. If this equilibrium between the EUF–N fractions does not exist, the EUF–NO$_3$ fraction cannot be used for the calculation of the quantity of easily available N. In fact, this quantity is underestimated in the case of leaching of NO$_3$ on account of the fact that easily soluble organic N compounds (Σ EUF–N – EUF–NO$_3$) from which NO$_3$ is formed by mineralization are not as easily removed by leaching as are the NO$_3$ ions. This quantity, if calculated by means of EUF–NO$_3$, is overestimated, when high amounts of NO$_3$ are added to the soil, for instance in the form of slurry, as the addition of slurry[3] does not increase the amount of EUF–N without NO$_3$. When the quantity of easily available N is calculated, it is, therefore, advisable to use the sum of EUF-extractable N and to base the calculation on 1 mg EUF–N/100 g soil being equivalent to 30 kg N/ha[12].

EUF-phosphorus fractions

Depending on the level of P concentration in the soil solution, about 5–10% of the lactate-P, as determined in many European soil testing laboratories, can be extracted by EUF. If soils differ widely in certain properties (exchangeable Al, CaCO$_3$ etc.), there is a poor correlation between the EUF–P and the lactate-P values. When the voltage is increased from 200 V to 400 V and the temperature raised from 20°C to 80°C, the amount of P extracted by EUF can be twice as high as that extracted at 200 V and 20°C.

For plant nutrition, however, not so much the quantity of extractable P but the quantity of available P is of primary importance. The easily available P can be

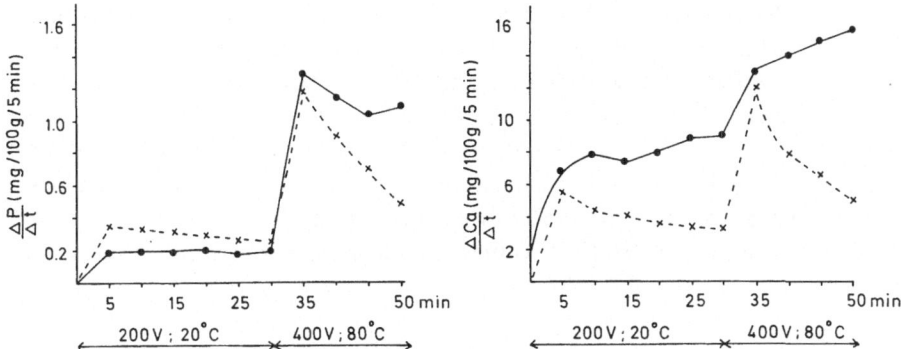

Fig. 3.   EUF–P extraction rates of two different soils at varying temperature and voltage.

obtained at 200 V and 20°C within 30 min[8,12]. The subsequent extraction at 400 V and 80°C gives information on the P reserves.

Fig. 3 shows that the calcareous pararendzina has a relatively low level of easily available P (200 V and 20°C) but high P reserves (400 V and 80°C). In contrast, the grey-brown luvisol is higher in easily available P, but the P reserves are lower than in the pararendzina. The lactate P_2O_5 values do not show these differences in P reserves (*cf* Fig. 3). The patterns of the EUF–P and EUF–Ca curves are very much alike. In the pararendzina, the nearly constant extraction rates at 20°C and 80°C indicate the presence of P reserves which will only be released at higher voltage and temperature, as is also found for Ca.

Correlations between EUF–P fractions and P-water values   The correlation between the P-values obtained by water extraction according to van der Paauw[9], and the EUF–P values at 200 V and 20°C during 30 min is very close[5]:

$y = 1.0 \times + 0.027; r = 0.97***; n = 109$
$y = $ P water values (mg/100 g)
$x = $ EUF–P (mg/100 g)

The P-water values in calcareous soils with high pH values are often lower than the EUF–P values at 20°C[8]:

$y = 0.26 \times + 0.26; r = 0.81***; n = 21$
$y = $ P-water values in calcareous soils (pH > 8.0)
$x = $ EUF–P

This difference is probably due to high Ca concentrations in the soil solution and high pH values which hamper P extraction by water. In the EUF procedure, on the contrary, Ca ions are continuously withdrawn from the soil solution, thus allowing continuous mobilization of phosphate ions[7].

EUF–P values required for optimal plant nutrition   Long-term field experiments have shown that the quantities of P that are easily available to a crop during a vegetation period can be well characterized by EUF extraction at 20°C and 200 V during 30 min[2,5,7,8,12]. These studies also showed that EUF–P value of 1.4–1.6 is necessary for optimal plant nutrition. The higher, however, the P reserves (EUF–P values at 80°C and 400 V within 30 to 35 min) are, the lower can be the required EUF–P values at 20°C. For calculation of the amounts of P fertilizer needed to raise the EUF–P values to 1.6 mg/100 g/30 min at 20°C, see Németh[6,7].

EUF potassium fractions

The course of the EUF–K curves at low and at high selectivity of the adsorption complex has already been thoroughly discussed in previous publications[4,5,6]. This paper will, therefore, concentrate on the relationship between EUF–K fractions and contents of exchangeable K (extracted with NH_4-acetate), as the content of exchangeable K is internationally used to characterize K availability.

The quantities of K desorbed within 30 min at 20°C and 200 V are generally lower than the contents of exchangeable K, i.e. they decrease with increasing contents of K-selective clay minerals in the soil and with decreasing degrees of K

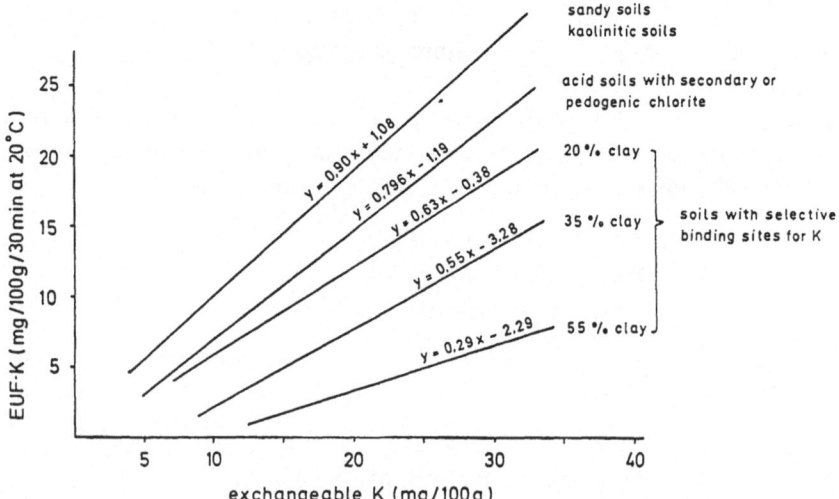

Fig. 4.   Correlations between EUF–K (mg/100 g/30 min at 20°C) and exchangeable K (mg/100 g) of different soils with different clay contents and different types of clay minerals.

saturation of these clay minerals (Fig. 4). In sandy soils and in kaolinitic soils, the contents of exchangeable K are nearly the same as the amounts of K obtained by EUF at 20°C and 200 V within 30 min, which can also be seen in Fig. 4. In acid soils where the sites which bind K selectively are occupied by Al and Al-hydroxy compounds, the differences between exchangeable K and EUF–K values are likewise small.

The correlation between exchangeable K contents and EUF–K values at 20°C can, therefore, only be significant, when clay content and type of clay are almost the same. The slopes of the regression lines (y = bx + a) clearly indicate in Fig. 4 that EUF–K values at 20°C (y) are lower than the values of exchangeable K (x) in soils having selective binding sites for K. At low K selectivity the slope is nearly 1.0. The higher the selectivity, the lower are these b values.

Correlations between K uptake, yield and EUF–K fractions   Long-term pot and field experiments show that the quantities of K which are easily available to a crop during a vegetation period can be well characterized by EUF extraction for a 30 min period at 20°C and 200 V [5, 8]. Statistically significant correlations were found between EUF–K fraction and K uptake as well as between this fraction and yield [2, 11, 12].

Fig. 5 shows an example with sugar beet as test crop. It can be clearly seen that sugar yield increases up to a K content of 15 mg/100 g/30 min at 20°C. This EUF–K value of 15 mg can be associated with exchangeable K contents (NH_4-acetate) of 15–45 mg/100 g depending on the soil properties. However this range becomes narrower, with decreasing strength of the solutions which are used for the extraction of K (0.025 M $CaCl_2$ or even water).

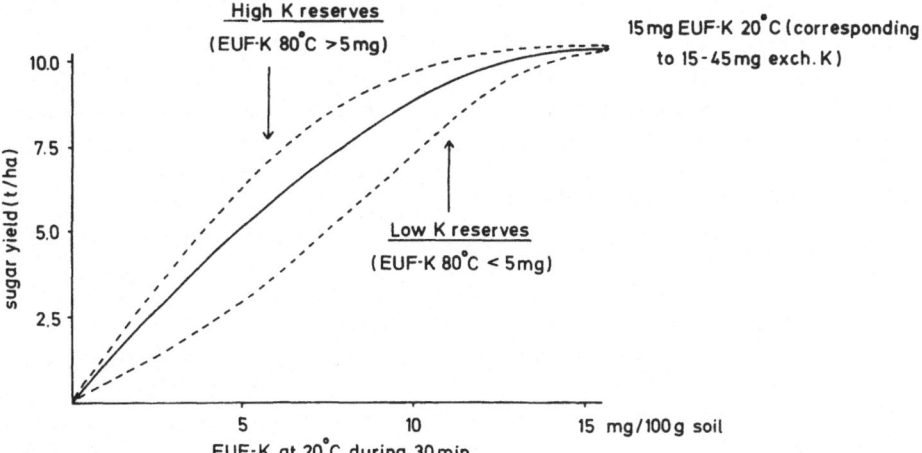

Fig. 5.   Relationship between EUF–K fractions and yield of sugar beet as established in long-term experiments.

Fig. 5 furthermore shows that sugar yields of 10 tons/ha can also be obtained at EUF–K values lower than 15 mg/100 g/30 min at 20°C. Such a variability indicates that the parameter of mg/100 g/30 min at 20°C does not satisfactorily predict the quantities of K that can be supplied to plant roots in the course of a vegetation period. Likewise, a 30-min extraction period at 20°C is not sufficient to characterize the K reserve (K-buffering capacities). The same applies to extraction with weak extractants like water or 0.025 M $CaCl_2$.

These K reserves can, however, be well characterized by means of EUF when extraction is carried out at 80°C and 400 V. For this purpose the extraction of 30 min at 20°C is continued for a further 5 min at 80°C and 400 V. The higher the K amounts obtained from 30 to 35 min at 80°C, the lower can be the EUF–K values at 20°C required for optimal plant nutrition (cf Fig. 5). For calculation of the amount of K fertilizer needed to raise the EUF–K value to 15 mg/100 g/30 min at 20°C, see Németh [5, 6].

These treshold values proved highly suitable for sugar beet crops at a yield potential of 50–55 t beets/ha [8, 12]. At a higher or a lower yield potential, these values have to be modified accordingly. The influence of differences in rooting depth on the required EUF–K values has also to be taken into account [10].

Determination of quantity of clay minerals by means of EUF–K fractionation   It is possible to determine the clay content of a soil by measuring the rate of anodical water flow [5], but the manufacturing of appropriate EUF filters allowing a constant water flow still poses some problems. It was, therefore, attempted to find other ways to determine the amount of clay minerals. As the speed of K desorption depends on the type and quantity of clay minerals, it was investigated, whether the clay minerals can be determined quantitatively with the aid of EUF–K fractions.

Fig. 6 shows that the EUF–K values at 80°C (F_2) increase with increasing clay content. This is the case for both low EUF–K values (a) and high EUF–K values (b). This means that at equal clay content the EUF–K values at 80°C increase with increasing K contents in the soil. The EUF–K values at 80°C are, therefore, not suitable as an index of the clay content.

Fig. 6 furthermore shows that the value $\dfrac{\text{EUF–K } 80°C}{\text{EUF–K } 20°C}$ also depends on the clay content. In sandy soils this value is rather low (Ia = 0.2). In a heavier soil (38% clay) this value is much higher (IIIa = 1.3) and it changes slowly with increasing K content (IIIb = 1.06).

When plotting the EUF–K values at 80°C (x) against $\dfrac{\text{EUF–K } 80°C}{\text{EUF–K } 20°C}$ (y), the relationship turns out to be negative (Fig. 7). The slopes of the regression depend on the clay contents, and an unknown clay content can, therefore, be estimated from these regression lines, when the values of EUF–K 20°C and EUF–K 80°C are known.

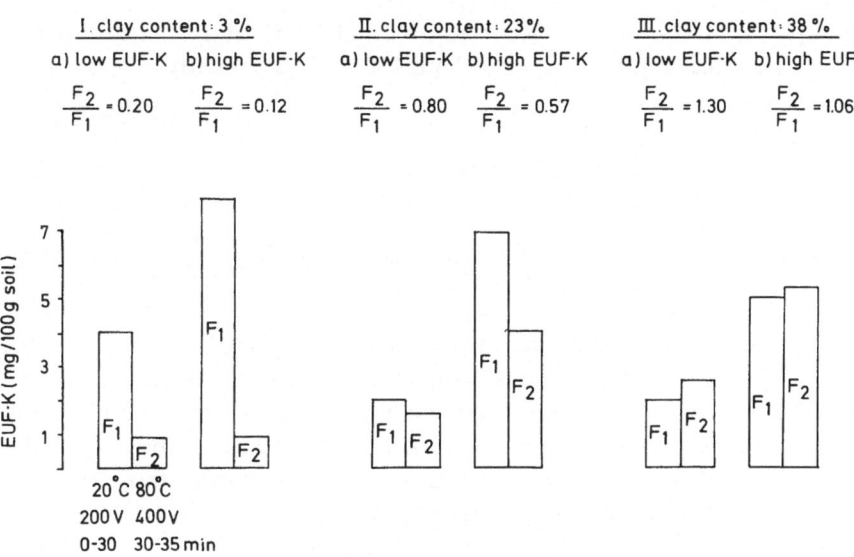

Fig. 6. Changes in the values of $\dfrac{\text{EUF–K } 80°C}{\text{EUF–K } 20°C}\left(\dfrac{F2}{F1}\right)$ with increasing amounts of EUF–K in the soils at different contents of K-selective clay minerals (3–38%).

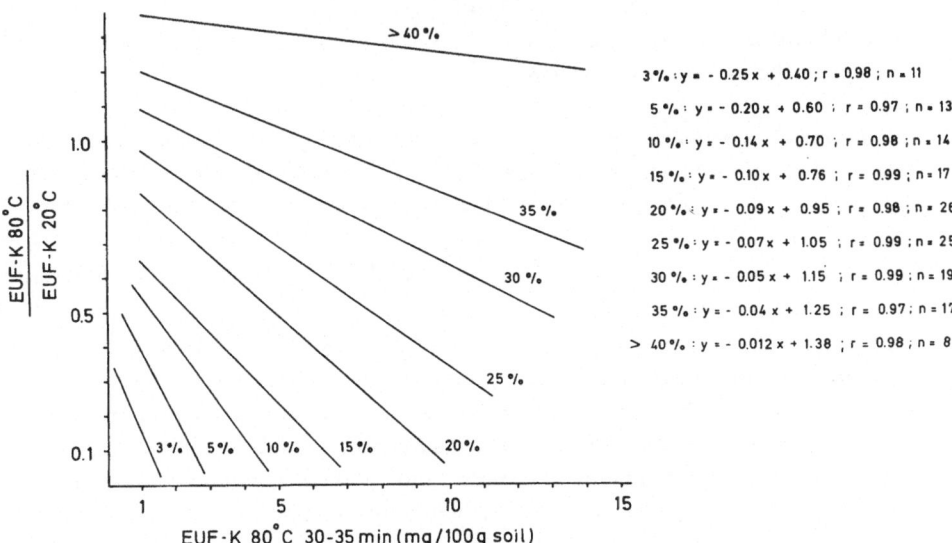

Fig. 7. Evaluation of the quantity of K-selective clay minerals by means of EUF–K fractionation.

Example a) A soil has an EUF–K value at 20°C of 13.9 mg/100 g and an EUF–K value at 80°C (x) of 6.7 mg/100 g. Consequently, the value of $\dfrac{\text{EUF–K 80°C}}{\text{EUF–K 20°C}}$ is 0.48 (y). At y = 0.48 and at x = 6.7, according to Figure 7, the clay content of this soil is approximately 20%.

Example b) A soil has an EUF–K value at 20°C of 9 mg and an EUF–K value at 80°C (x) of 3.0 mg/100 g. Consequently, the value of $\dfrac{\text{EUF–K 80°C}}{\text{EUF–K 20°C}}$ is 0.33 (y). At y = 0.33 and at x = 3.0, the clay content of this soil is approximately 10%.

Close correlations between the clay contents determined by EUF and those measured by conventional methods (sedimentation analysis) are, however, only found, when minerals of the clay fraction consist of illites, expanded illites and smectites with high charge. The EUF–K fractions do not assess the particle size distribution of a soil, but indirectly measure the quantities of those minerals which release K ions from their interlattice sites.

EUF heavy metal fractions

As the hydroxides and hydrated oxides of heavy metals are hardly soluble in the alkaline cathode filtrate, they remain in the cathode filter and cannot be obtained in fractions at 5-minute intervals. However, since it consists of soluble and desorbable ions, this EUF-heavy metal fraction is important for plant nutrition. Heavy-metal oxides of the soil do not migrate in the electric field. Therefore they cannot be extracted by EUF and have to be determined by conventional methods (extraction with acids).

Some of the heavy metals may become partly dissolved in the alkaline cathode filtrate in the form of complex compounds which will then be found in the cathode filtrate. This is the reason why the reproducibility of the heavy-metal values in the cathode filter is low.

Relationship between EUF-micronutrients and plant uptake Significant correlations could be found between EUF–Mn values and Mn contents in roots as well as in leaves of sugar beet[12]. This finding confirmed the results of previous investigations carried out with red clover[5].

Close correlations have also been observed between the EUF–Zn values and the Zn uptake of red clover, where the influence of K supply on Zn uptake becomes evident. Fig. 8 clearly shows that at a given EUF–Zn content (*e.g.* at 5.0 ppm) Zn uptake decreases with increasing K content in red clover.

Evaluation of anaerobic conditions in soils by means of EUF–Mn values If soils that are well supplied with Ca show high EUF–Mn values (< 2 ppm), this indicates that in these soils temporarily anaerobic conditions prevail which are mostly due to incorrect soil cultivation.

Fig. 9 gives the EUF–Mn contents and pH values in the profile of a grey-brown

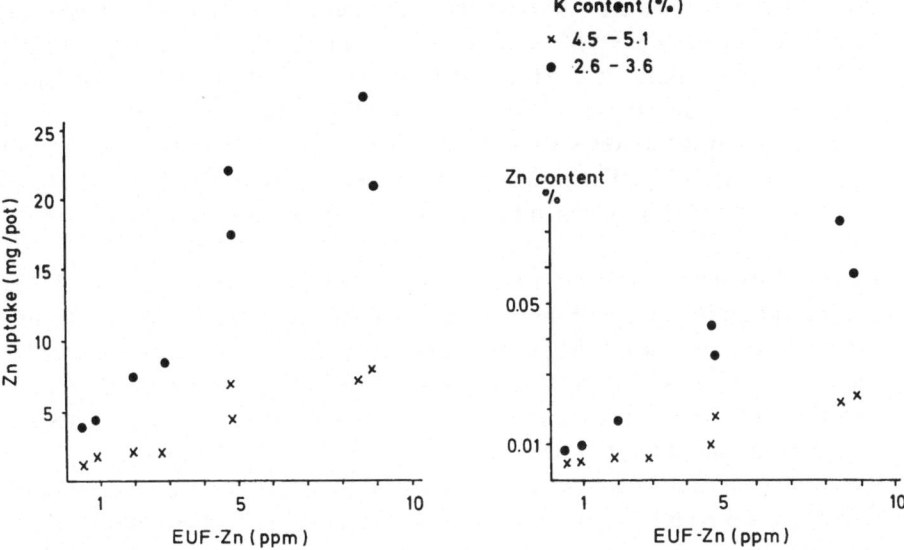

Fig. 8.   Relationship between EUF–Zn content in soil and Zn uptake and Zn content of red clover varying in K content.

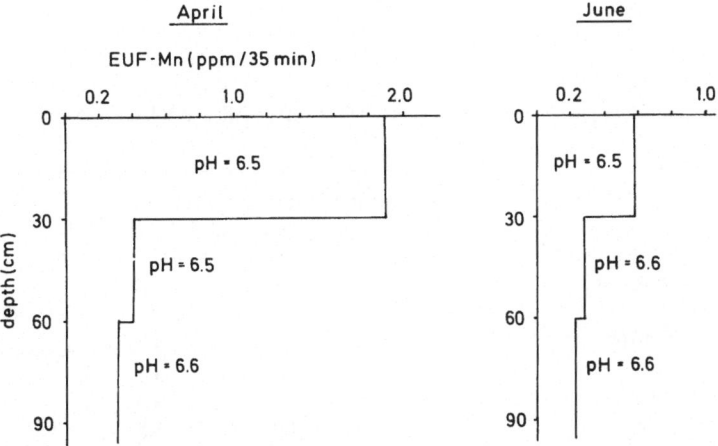

Fig. 9.   EUF–Mn values at various depths of a grey-brown luvisol having temporarily anaerobic conditions in the topsoil.

luvisol derived from loess. Although the pH values in the profile show practically no differences, the highest Mn values are found in the topsoil. Consequently, the high Mn values cannot be due to a higher proton concentration in the topsoil. In the year of sampling, the grey-brown luvisol of Fig. 9 was cultivated in the spring while the soil was too wet, so that anaerobic conditions prevailed in the topsoil and the content of EUF–Mn increased. The EUF–Mn contents in the topsoil clearly decreased (Fig. 9), when the soil was dry again (in June).

Calculation of physiological lime requirements   It is not so much the high H^+ ion concentration in soil solutions, but the toxic effects of Al, Fe, Mn or Zn ions which are responsible for the limited growth of plants on acid soils. The need to reduce the toxic concentrations of heavy metals is, therefore, of primary consideration, when the physiological lime requirements of a soil are assessed.

If the amount of Ca extracted within 30–35 min at 80°C and 400 V is below 2 mg/100 g soil, a soil is known to have very low Ca reserves. In such cases there is a risk of heavy metal toxicity, if the soil contains heavy metals by nature. In most soils the heavy metal contents increase with increasing clay content. It is, therefore, important to know the contents of easily desorbable heavy metals in soils high in clay content, if the Ca reserves are low. If the soil contains large amounts (> 5 ppm) of EUF-desorbable Al, Fe, Mn *etc.*, liming for plant physiological purposes is to be recommended.

If the heavy metal contents cannot be measured for an assessment of the CaO requirements, it is sufficient to know the Ca reserves at 80°C and the clay content. The recommended quantities of liming material to be applied are shown in Table 7. The data in this table make it clear that, in order to reach the required EUF–Ca value of 60 mg/100 g at 20°C within 30 min and of 20 mg/100 g at 80°C from 30 to 35 min[12], the quantities of CaO to be applied depend on the clay content of the soil.

Table 7. Assessment of physiological lime requirements (CaO in t/ha) by means of EUF–Ca fractions and clay contents

mg/100 g		CaO requirements, tons/ha, at various clay percentages		
EUF–Ca at 80°C (30–35 min)	EUF–Ca at 20°C (0–30 min)	10%	10–20%	20–30%
< 2.0	< 10.0	2.0	5.0	8.0
2– 5	10–15	1.0	3.0	4.0
5–10	20–30	0.5	2.0	3.0
10–15	30–40	–	1.0	1.5
15–20	50–55	–	0.5	1.0
> 20	> 60	–	–	–

References

1 Bechold H 1925 Elektro-Ultrafiltration. Z. Elektrochem. Angew. Phys. Chem. 31, 496–498.
2 Eifert J, Várnai M and Szöke L 1982 The application of EUF in grape production. Plant and Soil 64, 105–113.
3 Harrach T Németh K and Werner G 1982 Effect of soil properties and soil management on the EUF–N fractions in different soils at equal climatic conditions. Plant and Soil 64, 115–127.
4 Németh K 1972 The determination of desorption and solubility rates of nutrients in the soil by means of electro-ultrafiltration (EUF). Proc. 9th Colloq., Int. Potash Inst. 3–12.
5 Németh K 1976 Die effektive und potentielle Nährstoffverfügbarkeit im Boden und ihre Bestimmung mit Elektro-Ultrafiltration (EUF). Habilitationsschrift, Universität Gießen.
6 Németh K 1979 The availability of nutrients in the soil as determined by electro-ultrafiltration (EUF). Adv. Agron. 31, 155–187.
7 Németh K and Makhdum M I 1981 The evaluation of nutrient dynamics in calcareous soils from Pakistan by means of electro-ultrafiltration (EUF). Soil Sci. Plant Nutr. 27, 159–168.
8 Németh K und Wiklicky L 1980 Erfahrungen mit der EUF-Methode bei der Düngeberatung. Kali-Briefe Büntehof 15, 15–35.
9 Paauw F van der 1969 Entwicklung und Verwertung einer neuen Wasserextraktionsmethode für die Bestimmung der pflanzenaufnehmbaren Phosphorsäure. Landwirtsch. Forsch. 23, 102–109.
10 Rex M 1982 Effect of soil depth on EUF-values required for an optimal plant nutrition. Proc. Int. EUF Symp. II, 337–343, Budapest, Hungary.
11 Wanasuria S, Mengel K and De Datta S K 1981 Assessment of available potassium in wetland rice soils by means of the electro-ultrafiltration technique. Plant and Soil 59, 23–31.
12 Wiklicky L 1982 Application of the EUF procedure in sugar beet cultivation. Plant and Soil 64, 115–127.

Reproducibility of Ca, K, Na and P contents in the different EUF fractions

K. NÉMETH and H. RECKE

Büntehof Agricultural Research Station, Hannover, Federal Republic of Germany

Key words EUF–Ca EUF–K EUF–Na EUF–P Maximally desorbable EUF–K Reproducibility

Summary The reproducibility of EUF–Ca, EUF–K, EUF–Na and EUF–P in the different fractions (20°C, 200 V and 80°C, 400 V) as well as the reproducibility of the quantities of maximally desorbable K obtained by EUF and their half-time values are described. The calculations are based on the measurement of 288 individual values (EUF fractions) per nutrient. The following results were obtained:

1. The reproducibility of the EUF–K and EUF–Na contents at 20°C (0–30 min) as well as at 80°C (30–60 min) can be considered as good. The coefficient of variation for the EUF–K contents of all measurements is 3.3%. This value is improved to 2.0% when the sums of the EUF–K values obtained at 20°C and 80°C are combined for all soil samples. The CV for the Na contents of all measurements is 3.3%.

2. The reproducibility of the EUF–Ca contents is better for soils in which desorption processes prevail than for those in which both desorption and dissolution processes of Ca compounds ($CaCO_3$) occur. The CV for the EUF–Ca contents of all measurements is 5.7%.

3. The volumes of anode extracts have a decisive influence on the amounts of P extracted by EUF at constant voltage and temperature. The quantities of P extracted at 20°C and 200 V must, therefore, be corrected with the aid of an equation presented in Fig. 1. The CV of the corrected EUF–P contents of all measurements at 20°C is 7.0%. This value is improved to 2.8% when the sums of the EUF–P values at 20°C and 80°C are combined.

The reproducibility of the EUF–P contents at 80°C and 400 V is good. It appears to be unaffected by the extraction volumes. The CV of all P measurements at 80°C, without taking the extraction volumes into account, is 4.3%.

4. The reproducibility of the quantities of maximally desorbable K obtained by EUF and of their half-time values at 80°C and 400 V is very satisfactory. These values are, therefore, suitable for characterizing K reserves in soils.

5. Only small quantities of nutrients are involved in the undesired transport of anions into the cathode extracts and of cations into the anode extracts. To avoid any sources of error, pooling of the anode and cathode extracts before analysis is recommended.

Introduction

It is known from previous work that the amounts of EUF-extractable nutrients depend on the voltage applied, on the temperature in the soil suspension and on the amounts of extract transported by the water flow past the anode and cathode[2,3]. If these factors cannot be kept constant during the extraction, the reproducibility of the quantities of extracted nutrient is unsatisfactory.

The problem of keeping the temperature and field strength (V/cm) or strength of electric current (mA) constant has meanwhile been solved[3]. The EUF filters were also improved over the years, but their water permeability is still not homogeneous enough. The aim of the present experiment is, therefore, to investigate more precisely how the volumes of anode and cathode extracts influence the contents of the extracted nutrients. Moreover, attempts will be made to correct the errors caused by differences in volumes extracted. Finally it will be investigated, to what extent anions may be carried away cathodically by the water flow and to what extent the water flow may transport cations anodically.

Material and methods

The investigated samples belong to alluvial soils (Leine alluvium South Hannover) and to grey-brown luvisols (South Hannover). In order to study the influence of variations in nutrient contents on the reproducibility of EUF values, the samples were taken from depths of 0–30 and 30–60 cm.

The soils were first extracted by EUF at 200 V and 20°C for a duration of 30 min. Subsequently the voltage was increased to 400 V while the temperature was raised to 80°C. The same sample was extracted for another 30 min. The total extraction took 60 min. The anode and cathode extracts were collected separately at 5-min intervals and their volumes were determined. Next, the concentrations of Ca, K, Na and P both in the cathode and anode extracts were measured.

The extracts were obtained by filtration through cellulose triacetate filters.

Results

Reproducibility of EUF–K contents and of the volumes of cathode extracts in the different fractions

An effective way to demonstrate the influence of variations in amounts of nutrients in the soil on the reproducibility of the EUF contents is to evaluate the results of the topsoil measurements (0–30 cm) separately from those of the subsoil (30–60 cm).

The mean EUF–K contents of the soil samples taken from horizons at a depth of 0–30 cm and the mean volumes of cathode extracts are given in Table 1. These data clearly indicate that the reproducibility of the sums of the EUF contents at 20°C as well as those at 80°C is satisfactory, as the coefficients of variation (CV) are below 5% (1.3–4.3%). The reproducibility is even better, when the summations of the EUF–K of values at 20°C and 80°C are taken. In that case the CV values lie between 0.8–2.4%.

For the soil samples taken at 30–60 cm depth which have very low K contents, the reproducibility of the EUF–K contents is not as good as for the samples with higher K contents (cf Table 1). The respective CV values are in the range of 1.2–7.1%. The CV of 7.1% refers, however, to a very low mean EUF–K value of 1.6 mg/100 g soil/30 min. The measurement of these low quantities of extractable K is highly difficult and probably responsible for this large coefficient of variation.

As higher amounts of K are extracted at 80°C, the reproducibility of the EUF–

Table 1. Indices of reproducibility of quantities of EUF-extractable nutrients (mg/100 g soil) and of volumes of cathode extracts (ml). The values listed pertain to the summated quantities obtained over six 5-min intervals of a 30-min extraction period from 4 replicate samples of 2 layers (a = 0–30 cm; b = 30–60 cm) of 3 soils for 2 combinations of temperature and voltage (I = 20°C, 200 V; II = 80°C, 400 V; III = Σ(I + II))

		Reproducibility indices of												
		Nutrient content						Extract volume cathode						
		$\bar{x}$		$s_{\bar{x}}$		CV		$\bar{x}$		$s_{\bar{x}}$		CV		
		a	b	a	b	a	b	a	b	a	b	a	b	
K soil 1	I	4.4	1.5	0.08	0.04	1.8	2.5	58.7	59.8	3.7	3.1	6.3	5.2	
	II	11.2	5.9	0.15	0.07	1.3	1.2	65.5	66.2	3.8	2.6	5.8	3.9	
	III	15.6	7.4	0.13	0.09	0.8	1.2							
soil 2	I	9.2	3.5	0.22	0.18	2.4	5.1	77.6	65.0	7.3	10.9	9.4	16.7	
	II	10.8	6.7	0.46	0.40	4.3	5.9	95.7	103.0	3.7	8.4	3.9	8.2	
	III	20.0	10.2	0.28	0.36	1.4	3.5							
soil 3	I	4.4	1.6	0.17	0.11	3.9	7.1	68.0	65.4	9.2	3.7	13.5	5.6	
	II	12.6	4.0	0.27	0.07	2.1	1.7	76.6	118.1	7.2	5.4	9.4	4.6	
	III	17.0	5.6	0.41	0.14	2.4	2.5							
Ca soil 1	I	48.7	51.6	1.71	2.44	3.5	4.7							
	II	292.2	302.2	15.79	11.16	5.4	3.7							
	III	340.9	353.8	17.74	13.25	5.2	3.7							
soil 2	I	38.4	31.7	1.26	3.57	3.3	11.3							
	II	80.8	73.3	0.44	4.10	0.5	5.6							
	III	119.2	105.0	1.11	3.12	0.9	3.0							
soil 3	I	39.7	32.8	3.62	1.45	9.1	4.4							
	II	120.5	64.8	12.12	4.20	10.1	6.5							
	III	160.2	97.6	15.55	4.05	9.7	4.2							

K values is better at 80°C than at 20°C. The reproducibility of the values is further improved, when the sums of the EUF–K values at 20°C and 80°C are combined. The respective coefficients of variation lie between 1.2 and 3.5% (cf Table 1).

When the coefficient of variation for the EUF contents of all investigated soil samples and both EUF treatments (20°C and 80°C) is calculated, the value is 3.3% which can be regarded as very good. This value decreases to 2.0%, when the sums of the EUF–K values at 20°C and 80°C are combined.

These results can be considered as reliable, for the coefficients of variation were calculated on the basis of 288 individual measurements (6 soils × 2 EUF treatments (20°C and 80°C) × 6 fractions × 4 parallel determinations).

The variations in volume of cathode extracts are higher than the variations in EUF–K contents. Twice CV values of more than 10% (13.5 and 16.7) were found in the cathode extracts. The relatively large variations in extraction volumes from sample to sample did, however, not influence the reproducibility of the EUF–K values, when extraction volumes of more than 50 ml/30 min were concerned. Extraction volumes above 100 ml/30 min, on the other hand, do not have any influence on the EUF-K contents. Consequently it is necessary to make sure that the volumes of the cathode extracts are not below 50 ml/30 min.

Reproducibility of EUF–Ca and EUF–Na contents

The reproducibility of the EUF–Ca contents in soils where mainly Ca desorption processes take place is very good at 80°C. A rapid decline in EUF–Ca desorption rates at 80°C during the extraction indicates the occurrence of desorption processes. The EUF–Ca values at 80°C fall off slowly or not at all when soils are concerned in which Ca undergoes both desorption and dissolution. As the rate of dissolution of the Ca compounds depends on the particle size, the reproducibility of the EUF–Ca contents can be influenced by particle size and distribution of the Ca compounds.

The volumes of cathode extracts also influence the reproducibility of the EUF–Ca amounts, if these volumes are below 50 ml/30 min. In one case only 22.3 mg Ca were extracted at a volume of 33.8 ml, whereas 30–37 mg Ca were measured for extraction volumes above 50 ml. Accordingly, there was a delayed extraction of Ca ions at extraction volumes below 50 ml, so that these Ca amounts were later recovered in the 80°C fractions. The reproducibility of the EUF–Ca amounts is therefore improved when the EUF–Ca amounts at 20°C and 80°C are pooled (*cf* Table 1).

When the coefficient of variation for the EUF–Ca amounts of all investigated soils and both EUF treatments (20°C and 80°C) us calculated, a value of 5.7% is obtained. This value is improved to 4.5% by pooling the sums of the EUF–Ca values at 20°C and 80°C.

Since, contrary to Ca, Na does not form difficultly soluble compounds in the soil, the reproducibility of the EUF–Na contents can be considered as good. Moreover, the Na contents in the soil are lower than the Ca contents. Consequently, small extraction volumes are already sufficient to transfer the desorbed Na ions from the electrode into the extracts.

When the coefficient of variation for the EUF–Na contents of all investigated soils and both EUF treatments (20°C and 80°C) is calculated, the value is 3.3% which can be improved to 3.0% by pooling the sums of the EUF–Na contents at 20°C and 80°C.

Reproducibility of EUF–P contents and of the volumes of anode extracts in the different EUF fractions

The amounts of P extracted by EUF are considerably influenced by the volumes of anode extracts. This can be explained by taking into account the accumulation at the anode of negatively charged soil particles obstructing the transfer of phosphate ions into the anode extracts. Consequently, the quantity of P extracted at equal P concentration in the EUF middle cell will decrease with decreasing volume of the anode extracts[3]. For example, 3.3 mg P/100 g are obtained at an extraction volume of 66 ml/30 min (20°C), whereas only 1.3 mg P/100 g are transferred into the anode extracts when the extraction volume is only 21 ml. In the 80°C treatment, however, where the extraction volumes exceed 65 ml/30 min, the quantities of P extracted are almost equal.

To avoid variations in EUF–P contents caused by different extraction volumes it is suggested to adjust the quantities of P extracted to an extraction volume of 65 ml. This correction is based on a quadratic regression between the quantities of P extracted and the volumes of anode extracts[3]. Correction of the P values is not necessary, when the extraction volumes are larger than 65 ml/30 min which normally is the case for the EUF treatment at 80°C. It can be seen that the reproducibility of the EUF–P contents is better at 80°C than at 20°C. The best reproducibility is, however, obtained when the sums of the P contents at 20°C and 80°C are combined. The explanation for this improvement is that P, which was already extracted at 20°C but which had been transported past the anode because of small extraction volumes, was transferred rapidly at 80°C.

When the coefficient of variation for the sums of all P measurements is calculated, a value of 5.7% is obtained which is not quite satisfactory. This value, however, improves to 3.6% when the sums of P at 20°C and 80°C are combined.

The reason for the relatively high variation in EUF–P values is probably to be sought in extraction volumes being often below 25 ml/30 min. At these volumes

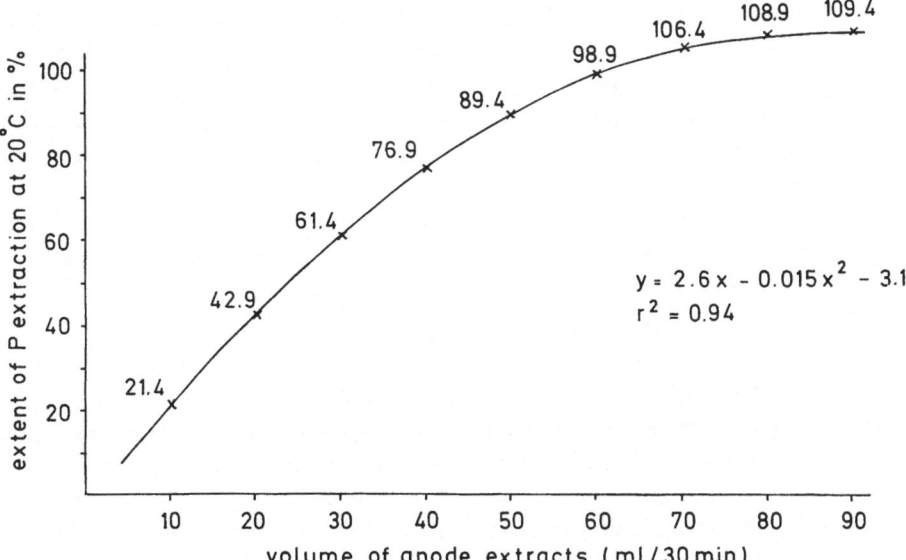

Fig. 1.   Influence of anode extract volumes on the amounts of P extracted by EUF at 20°C.

Table 2. Indices of reproducibility of quantities of EUF-extractable P (mg/100 g soil) and of volumes of anode extracts (ml). The values listed pertain to the summated quantities obtained over six 5-min intervals of a 30-min extraction period from 4 replicate samples of 2 layers (a = 0–30 cm; b = 30–60 cm) of 3 soils for 2 combinations of temperature and voltage (I = 20°C, 200 V; II = 80°C, 400 V; III = Σ(I + II))

							Reproducibility indices of					
		P content						Extract volume anode				
		$\bar{x}$		$s_{\bar{x}}$		CV		$\bar{x}$		$s_{\bar{x}}$		CV
		a	b	a	b	a	b	a	b	a	b	a	b
soil 1 I		0.97	0.21	0.02	0.01	2.6	5.7	22.2	44.7	2.0	18.8	9.0	42.1
II		3.83	1.58	0.09	0.10	2.4	6.1	30.8	43.9	3.1	16.6	10.1	37.9
III		4.80	1.79	0.11	0.10	2.2	5.7						
soil 2 I		1.68	0.82	0.13	0.07	7.5	8.3	37.5	65.2	6.9	18.5	18.4	28.3
II		4.64	2.09	0.24	0.10	5.1	4.8	92.4	75.5	7.3	9.3	7.9	12.3
III		6.32	2.91	0.14	0.10	2.3	3.6						
soil 3 I		3.07	0.39	0.13	0.05	4.2	13.7	35.6	25.2	10.5	1.2	29.6	4.8
II		6.93	1.03	0.27	0.04	3.9	3.6	74.1	71.1	2.4	11.0	3.2	15.5
III		10.00	1.42	0.26	0.07	2.6	4.9						

the EUF–P values cannot be satisfactorily corrected to 65 ml with the formerly applied correction equation[3], because extraction volumes exceeding 30 ml/30 min were used when the relation between the extraction volumes and the extracted P amounts was worked out. As volumes below 25 ml/30 min were also obtained in the present investigation, the relation between quantities of P extracted and the extraction volumes had to be recalculated. The resulting quadratic regression equation is presented in Fig. 1. With the aid of this equation even those EUF–P contents can be corrected which are obtained at extraction volumes of less than 25 ml (Table 2).

If the EUF–P contents at 20°C are corrected, the coefficient of variation for all P measurements at 20°C is 7.0%. The CV calculated with the old correction is 11.3%. Consequently the variability of the EUF–P values at 20°C is considerably lower, when the correction according to Fig. 1 is applied.

As already described, the reproducibility of the EUF–P contents at 80°C and 400 V is very good. The CV for all P measurements at 80°C corrected with the aid of Fig. 1 is 4.5%. When the variation in EUF–P values at 80°C is calculated, the CV value is likewise found to be satisfactory, namely 4.3%. Consequently, a correction of the EUF–P values at 80°C is unnecessary. The reason is that the transfer of the phosphate ions at 80°C and 400 V does not depend as much on extraction volume as on temperature and voltage. At 20°C and 200 V, on the contrary, the extraction volume plays an important role in the extraction of P.

The EUF–P values must, therefore, be corrected with the aid of the equation presented in Fig. 1.

The reproducibility of the volumes of anode extracts was found to be still unsatisfactory, although the vacuum had been kept constant throughout the investigation. Consequently, it is believed that the variations in extraction volume are due mainly to faulty filters. The quantity of P extracted with the use of these filters can, however, be corrected with the aid of the equation given in Fig. 1.

Undesired transport of cations into the anode extract and of anions into the cathode extract

During electro-ultrafiltration the anions migrate to the anode and the cations move to the cathode. The separation of the ions in the electric field is influenced by the vacuum in the external EUF compartments which removes the water from the middle cell by suction. It is, therefore, quite well possible that the water flow may also transport cations into the anode extracts and anions into the cathode extracts. This irregularity may have an influence on the reproducibility of the EUF values, if the anode and cathode extracts are not pooled before analysis. The aim of our investigation was, therefore, to determine the percentages of the desorbed quantities of K and Ca which were transported into the anode extracts as well as of the desorbed quantities of P delivered to the cathode extracts. This determination was made for extractions at both 20°C and 80°C.

At 20°C it was found that 2–5% of the desorbed K and only 1–2% of the desorbed Ca can be carried away by the water flow into the anode extracts, and that 20–30% thereof were already transferred within the first 5 min. After 5 min the transfer slows down, because clay minerals accumulating at the anode filter prevent the further transport of cations into the anode extracts.

At 80°C less than 2% of the total desorbed K and less than 1% of the total desorbed Ca were transported by the water flow into the anode extracts, so that they may be neglected.

The quantities of P delivered to the cathode extracts at 20°C are below 2% of the total P desorbed. These 2% are almost completely transferred within the first 5 min and can, therefore, be neglected.

The degree of irregularity depends on the difference between the volumes of the anode and cathode extracts. For instance, the quantities of cations transported into the anode extracts increase with increasing volumes of anode extracts relative to those of cathode extracts, and vice versa. Moreover, the degree of irregularity increases with increasing concentration of the respective nutrients in the middle cell of the EUF apparatus. Pooling the anode and cathode extracts for subsequent analysis has, therefore, proved highly suitable in routine investigations.

Reproducibility of the maximally desorbable quantities of K obtained by EUF and their half-time values

As the individual EUF–K fractions are often used for the calculation of maximally desorbable (d_m) K [1,3], it is important to evaluate the reproducibility of each EUF fraction with 5-min intervals separately. If the variability in each EUF fraction is high, there is reason to believe that the reproducibility of the values of maximally desorbable K and of their half-time values $(t_{0.5})$ is unsatisfactory.

The calculation of the quantities of maximally desorbable K at 20°C and 200 V poses problems in many soils in agricultural use, as the desorption rates decrease only slightly in the course of an extraction. At 80°C and 400 V the quantity of maximally desorbable K can be calculated without difficulty, as the desorption rates of the topsoil as well as of the subsoil decrease rapidly. The reproducibility of the values of maximally desorbable K and of their half-time values at 80°C is found to be good. In the subsoil, however, where the d values (mg/100 g/30 min) are low, the variations in half-time values can be slightly higher. The d_m values and the $t_{0.5}$ values at 80°C and 400 V are, however, very well suited to characterize the K reserves.

As already mentioned, it is often not possible to calculate the d_m and $t_{0.5}$ values at 20°C and 200 V, because the desorption rates hardly change during extraction. If such values can be determined at all, it is to be expected that their variability is rather high, in particular when the d values are low in comparison with the d_m values and the half-time values are correspondingly long. Despite very similar d values, the d_m values and their half-time values can be different, when the change in desorption rate with time varies too much among replicates. The explanation for these variations in rates of desorption is that measuring quantities of K below 0.7 mg/100 g/5 min is difficult. It is, therefore, recommended to calculate the d_m values only in those cases in which the desorption rates can be measured with more precision. At 80°C and 400 V, on the contrary, these parameters can be adequately calculated and used for the characterization of K reserves.

Reproducibility of EUF nutrient contents in routine analysis

For fertilizer recommendations in agricultural practice it is generally sufficient to work with two EUF fractions, namely with the fraction at 20°C and 200 V during 0–30 min and with the fraction at 80°C and 400 V during 30–35 min. The reproducibility of the EUF nutrient fraction at 20°C was already discussed in detail. Our attention will, therefore, be focussed on the reproducibility of the EUF–Ca, EUF–K, EUF–Na and EUF–P contents extracted at 80°C within 30–35 min.

The reproducibility of these values is satisfactory in all samples investigated. The coefficients of variation of all values are 6.0 for Ca, 3.7 for K and 4.6 for Na. These EUF values obtained within 30–35 min lend themselves well for characterizing the Ca and K reserves. There appears to be no need to extend the extraction for another 5 min. This reproducibility of the EUF–P values at 80°C within 30–35 min is not as good as for the EUF–Ca, EUF–K and EUF–Na values. The CV of 8.3% calculated for all samples investigated can, however, be considered as satisfactory. This value is relatively high due to the fact that the quantities of P in this EUF fraction are generally low. It has, moreover, to be taken into account that a delayed transfer of the phosphate ions at 20°C due to low extraction volumes may influence the P contents of the EUF fraction at 80°C during 30–35 min. P which had been extracted but was not yet transferred at 20°C may increase the P contents in the first EUF fraction at 80°C. This assumption is corroborated by the finding that the reproducibility of the combined EUF–P values at 20°C and 80°C is better than when the results obtained at 20°C and 80°C are calculated separately. When P contents at 20°C and 80°C are combined, the variability caused by a delayed transfer is decreased. As the EUF–P contents required for optimal plant nutrition are expressed by a combination of EUF–P values at 20°C and 80°C, the reproducibility of the sums of P at 20°C and 80°C plays a decisive role in the procedure of making fertilizer recommendations. The CV for the combined EUF–P contents at 20°C and 80°C of all soil samples is 4.4% which can be considered as satisfactory.

References

1 Grimme H 1979 The use of rate equations for a quantitative description of K desorption from soils in an external electric field (electro-ultrafiltration). Z. Pflanzenernaehr. Bodenk. 142, 57–68.
2 Németh K 1972 Bodenuntersuchung mittels Elektro-Ultrafiltration (EUF) mit mehrfach variierter Spannung. Landwirtsch. Forsch. Sonderh. 27, 184–196.
3 Németh K 1976 Die effektive und potentielle Nährstoffverfügbarkeit im Boden und ihre Bestimmung mittels Elektro-Ultrafiltration. Habil. Schrift, Universität Gießen.

Test of the electro-ultrafiltration method's applicability in soil analysis. Reproducibility of the method

G. SIMÁN

Dept. of Soil Sciences, Swedish University of Agricultural Sciences, S-750 07 Uppsala 7, Sweden

Key words   Electro-Ultra-Filtration   Soil analyses

Summary   The applicability of the Electro-Ultra-Filtration (EUF) method in soil analyses was studied. The reproducibilities of the amounts of soil extracts, of ion concentrations in the extracts and of the distribution of cations and anions over the cathode and anode extracts by use of fully automatic EUF equipment were tested.

The degree of variability among replicates was expressed as coefficient of variation (CV) and as the highest percentual divergence of an individual analytical measurement from the mean (L).

The extraction volumes of five replicates of six different soils were found to vary between 1.1–7.1% with an average of 3.8%, as CV and between 1.5–11.3% as L. The reproducibility of desorbed P in the anode extract varied between 2.7–31.7% with an average of 8.7%, as CV and between 3.2–37.9% as L. Corresponding values for CV and L of K desorbed varied between 1.3–13.9% and 1.6–23.8%, respectively.

Variations among replicates of desorbed P were especially high in the first 1–2 sub-fractions of a total of seven fractions in a single extraction run. Low K concentrations in the extract had a slightly negative influence on the reproducibility of K desorption.

Furthermore, it was found that a portion of the cations is collected in the anode extract and a portion of the anions in the cathode extract, especially at the beginning of an extraction run. Pooling of anode and cathode extracts before analysis is therefore recommended.

Introduction

The possibility of determining the amounts of effective available and exchangeable nutrients and the desorption rates of exchangeable nutrients in the soil by use of the Electro-Ultra-Filtration method (EUF) according to Németh[1] has been noted with great interest by many soil scientists. Knowledge of these soil characteristics provides a possibility to consider the dynamic nature of plant nutrients that is so essential for improved regulation of the nutrient supply to the cultivated crops.

Following the acquisition of EUF equipment by the Dept. of Soil Sciences plans were made to test the EUF method's applicability in soil analyses and fertilizer recommendations. As a first step the precision of the method was investigated. EUF extraction removes only certain fractions of the total quantities of nutrients in the soil. Because of the incompleteness of the extraction the extracted quantities of nutrients are easily affected by slight variations in the extraction procedure. Therefore the reproducibility of the EUF-fractions is a very important aspect of the method.

The present paper deals with the matter of reproducibility of the amount of extracts, the reproducibility of nutrient concentrations in the extracts and the distribution of nutrients in the cathode and anode extracts.

Material and methods

The soil samples in this investigation were collected from six long-term soil fertility experiments situated in the county of Malmöhus in southern Sweden. The plots from which the soil samples were taken had been fertilized annually since the start of the trial in 1958 with 15 kg/ha P and 40 kg/ha K in addition to compensation for removed P and K in the harvested crops. The soil samples were taken in the autumn of 1977. Some basic data on the experimental soils are given in Table 1.

A fully automatic EUF equipment, model 723, was used for extraction of the soil samples. The total extraction took 35 min. divided into 5 min. intervals. The voltages between the electrodes was 50 V for 0–5 min., 200 V for 5–30 min. and 400 V for 30–35 min. in agreement with the recommendations of Németh[1,2]. Anode and cathode filters were of the types EUF 510 and EUF 511, respectively. The anode and cathode extracts were collected separately. The vacuum in the two outside chambers was adjusted so as to allow 50-ml volumes of the cathode extract (soil extract + rinse water) per 5 min. extraction time to be obtained.

Five replicate samples of each soil type were extracted. The concentrations of K and P in both cathode and anode extracts were determined. K was determined on the flame emission spectrometer and P colorimetrically. The reproducibility of the EUF method is given both as coefficient of variation, CV (standard deviation as per cent of the mean value) and as highest devergence of an individual analytical value from the mean value in per cent, L.

Results and discussion

Table 2 shows the amounts of the anode extracts and their degrees of variability. The investigation comprised six soils. During the 35 min. extraction a total of seven sub-fractions were collected. Each mean value in the table represents 5 parallel extractions of the same soil. The table shows that the volume

Table 1. Basic characteristics of the soils (0–20 cm depth) used in the investigation

Location	Mechanical composition, %					Chemical properties		
	Clay	Silt	Fine sand	Coarse sand	Org. C	pH (H$_2$O)	P–AL mg/100 g	K–AL mg/100 g
1. Fjärdingslöv	17.3	13.0	37.1	28.5	1.8	7.2	9.2	7.9
2. Orup	12.7	9.4	38.0	34.0	2.6	5.9	5.7	13.1
3. Västraby	29.0	17.0	20.1	28.9	2.3	6.5	8.2	13.1
4. Örja	15.2	16.0	41.7	24.0	1.3	7.4	8.4	15.8
5. S. Ugglarp	8.1	7.8	41.6	39.3	1.7	5.8	7.4	10.9
6. Ekebo	13.9	15.0	37.0	28.5	3.5	6.1	8.8	22.1

of sub-extract per 5 min. decreased with time depending on the quantity of clay deposited on the anode filter which reduced its permeability.

The degree of scattering of the individual determinations around the mean is usually expressed in the form of a standard deviation (SD). However, since SD is an absolute index and varies with the volume of the extracts, the coefficient of variation (CV) was calculated instead to enable a direct comparison of the reproducibility of parallel extractions between different soils and sub-fractions to be made. In addition, the highest deviation of an individual volume of extraction solution from the mean is given in order to provide a more stringent measure of the method's reproducibility. This divergence is also given as a percentage of the mean value (L).

The variability in the extraction volumes among replicates, expressed in CV, ranged between 1.1–7.1% and was, on average 3.8%. The highest divergence of an individual analytical volume from the mean value varied in the entire

Table 2. The amounts and the indexes of reproducibility of anode extracts. Mean value (x), coefficient of variation (CV) and the highest deviation of an individual volume of extract from the mean (L)

Extraction classes in minutes	1. Fjärdingslöv			2. Orup			3. Västraby		
	x ml	CV %	L %	x ml	CV %	L %	x ml	CV %	L %
0– 5	50.2	6.9	11.3	48.2	5.2	7.7	44.1	3.8	6.6
5–10	39.1	5.9	9.2	41.0	5.3	5.9	36.2	3.7	5.2
10–15	36.3	2.8	3.3	39.6	5.6	8.1	33.8	3.6	5.3
15–20	34.2	3.7	4.7	37.4	4.1	5.9	31.8	3.0	4.7
20–25	32.9	6.3	8.2	35.9	4.6	7.0	31.8	4.6	6.6
25–30	31.4	3.9	4.8	36.0	3.4	5.8	30.3	1.8	2.6
30–35	30.8	2.9	5.2	35.6	5.0	7.9	30.3	2.4	3.6

Extraction classes in minutes	4. Örja			5. S. Ugglarp			6. Ekebo		
	x ml	CV %	L %	x ml	CV %	L %	x ml	CV %	L %
0– 5	38.9	2.8	4.6	43.9	3.1	5.2	42.8	6.7	8.4
5–10	34.4	1.3	1.7	38.2	5.9	7.6	37.5	7.1	6.9
10–15	32.6	2.5	4.3	36.1	4.9	7.2	35.0	4.7	6.0
15–20	31.2	1.9	2.6	34.6	3.2	4.6	33.5	1.1	1.5
20–25	30.2	1.7	2.3	34.0	3.4	4.7	33.1	3.7	5.4
25–30	29.5	2.3	3.4	33.2	2.7	3.6	32.7	3.1	5.5
30–35	29.1	1.8	3.1	33.0	3.1	3.6	32.8	2.7	4.3

investigation material between 1.5 and 11.3%. In order to provide a comparison, it may be mentioned that Németh[2] found the highest deviation of an individual extraction volume from the mean to vary between 5.3 and 18.4%.

The rates of P desorption and their reproducibilities in six soils are given in Table 3. Each value of P concentration in the table is a mean of 5 parallel determinations. The tables shows that the reproducibility, expressed in CV, varied between 2.7–31.7% and was on average 8.7%. The highest deviations of individual P analysis values from the mean were found to vary between 3.2 and 37.9%.

The reproducibility of P desorption was not as good in the first subfractions as in the subsequent ones, which is demonstrated by the calculated linear

Table 3. The rates of phosphorus desorption in different soils and their reproducibilities with the use of the EUF method

Extraction classes in minutes	1. Fjärdingslöv			2. Orup			3. Västraby		
	x mg/ 100 g	CV %	L %	x mg/ 100 g	CV %	L %	x mg/ 100 g	CV %	L %
0– 5	0.27	11.6	19.4	0.04	10.1	13.0	0.14	4.2	7.3
5–10	0.33	8.0	12.1	0.09	7.6	11.1	0.19	4.9	6.2
10–15	0.26	6.3	8.4	0.09	7.6	11.6	0.16	5.4	6.6
15–20	0.20	6.6	9.1	0.08	7.9	10.5	0.13	6.7	9.4
20–25	0.16	6.7	11.1	0.08	8.0	10.1	0.11	7.3	10.5
25–30	0.13	4.1	4.5	0.06	7.0	9.4	0.10	5.7	9.3
30–35	0.18	6.5	9.1	0.08	6.5	10.3	0.12	2.7	3.4

Extraction classes in minutes	4. Örja			5. S. Ugglarp			6. Ekebo		
	x mg/ 100 g	CV %	L %	x mg/ 100 g	CV %	L %	x mg/ 100 g	CV %	L %
0– 5	0.19	7.7	12.9	0.20	13.2	23.2	0.08	31.7	37.9
5–10	0.28	2.9	3.2	0.26	17.6	20.8	0.15	27.9	35.8
10–15	0.23	7.5	10.4	0.21	17.4	27.4	0.13	16.6	20.6
15–20	0.19	6.4	10.6	0.17	10.9	12.1	0.10	6.4	8.7
20–25	0.16	5.7	8.3	0.14	6.5	10.7	0.09	7.7	11.5
25–30	0.14	6.3	8.3	0.13	3.6	6.3	0.09	6.5	10.3
30–35	0.18	4.8	6.9	0.16	11.3	17.9	0.10	6.0	7.5

relationship between the fraction numbers and CV, $Y = 13.77 - 1.26X$, where X-fraction number and $Y = CV$. Although the relationship is weak ($r = 0.43$) the trend is clear, namely that CV decreases from the first fraction to the seventh. On the other hand, the reproducibility of P desorption did not seem to be particularly influenced by the variation in P concentration in the extract.

Table 4 shows the amounts of K desorbed and their degrees of variability. The reproducibility expressed in CV varied between 1.3–13.9% with an average of 8.0%. The highest deviation of an individual analytical value from the mean (L) varied throughout the entire material between 1.6 and 23.8%. The reproducibility of K in the sub-fractions remained nearly the same throughout the period of extraction. On the other hand, low K concentrations in the extract had a slight negative influence on the reproducibility ($Y = 8.6 - 0.62X; r = -0.15$), where X

Table 4. The rates of potassium desorption in different soils and their reproducibilities with the use of the EUF method

Extraction classes in minutes	1. Fjärdingslöv			2. Orup			3. Västraby		
	x mg/ 100 g	CV %	L %	x mg/ 100 g	CV %	L %	x mg/ 100 g	CV %	L %
0– 5	0.26	7.0	9.4	1.35	1.3	1.6	0.98	6.7	9.0
5–10	0.38	11.9	16.4	2.08	4.7	5.6	1.60	11.5	17.9
10–15	0.32	9.0	14.9	1.38	5.0	5.8	1.21	9.9	14.0
15–20	0.26	5.7	8.4	0.87	3.9	6.7	0.92	4.9	6.9
20–25	0.24	10.1	13.4	0.61	5.9	8.5	0.79	7.4	11.2
25–30	0.22	10.1	14.7	0.42	12.1	15.8	0.68	4.1	5.0
30–35	0.30	6.1	8.1	0.53	5.1	7.2	0.80	6.6	8.8

Extraction classes in minutes	4. Örja			5. S. Ugglarp			6. Ekebo		
	x mg/ 100 g	CV %	L %	x mg/ 100 g	CV %	L %	x mg/ 100 g	CV %	L %
0–5	0.72	8.3	14.5	1.32	10.3	12.0	2.89	7.1	10.0
5–10	1.21	7.0	9.9	1.78	4.4	6.9	3.77	8.3	13.7
10–15	1.15	5.5	8.1	1.21	7.5	12.4	2.33	6.9	11.0
15–20	0.99	7.5	9.1	0.92	6.4	8.3	1.50	10.9	14.9
20–25	0.87	6.2	8.3	0.71	12.6	16.6	1.05	13.8	17.5
25–30	0.76	11.5	15.8	0.56	11.7	14.1	0.84	7.0	10.5
30–35	1.02	13.9	23.8	0.67	12.8	18.7	1.01	6.3	8.3

= K conc. in the extract and Y = CV. The reproducibility expressed both in CV and in L, was slightly better for K then for P.

In order to provide a means of comparison, it can be mentioned that in Sweden the following scale is used for the evaluation of the reproducibility of soil analyses expressed as the highest deviation from the mean of parallel analyses (Karlsson, pers. comm.) 0–5.0% good, 5.1–10.0% fairly good, 10.1–20.0% large deviation, but acceptable and > 20.1% generally unacceptable.

Detailed examination of the present EUF material reveals that in the case of K there was only one replicated determination that exceeded the unacceptable level of 20.0%, while for P there were two soils (nos. 5 and 6) at the start of the extraction.

Table 5. Distribution of desorbed potassium over cathode (C) and anode (A) extracts. Mean values of five parallel extractions

Extraction classes in minutes	1. Fjärdingslöv		2. Orup		3. Västraby	
	K in C + A mg/100 g	Percent K in A	K in C + A mg/100 g	Percent K in A	K in C + A mg/100 g	Percent K in A
0– 5	0.44	40.9	1.80	25.0	1.15	14.8
5–10	0.51	25.5	2.32	10.3	1.67	4.2
10–15	0.44	27.3	1.47	6.1	1.25	3.2
15–20	0.41	36.6	0.91	4.4	0.95	3.2
20–25	0.33	27.3	0.66	7.6	0.84	6.0
25–30	0.30	26.7	0.46	8.7	0.71	4.2
30–35	0.38	21.0	0.57	7.0	0.80	0.0
0–35	2.82	30.1	8.19	11.6	7.37	5.3

Extraction classes in minutes	4. Örja		5. S. Ugglarp		6. Ekebo	
	K in C + A mg/100 g	Percent K in A	K in C + A mg/100 g	Percent K in A	K in C + A mg/100 g	Percent K in A
0– 5	0.84	14.3	1.55	14.8	3.27	11.6
5–10	1.25	3.2	1.89	5.8	3.91	3.6
10–15	1.19	3.4	1.26	4.0	2.42	3.7
15–20	1.03	3.9	0.95	3.2	1.54	2.6
20–25	0.92	5.4	0.73	2.7	1.07	1.9
25–30	0.81	6.2	0.58	3.4	0.86	2.3
30–35	1.06	3.8	0.70	4.3	1.05	3.8
0–35	7.10	5.4	7.66	6.4	14.12	5.2

Another important question is whether it is sufficient to determine anions only in the anode extract and cations only in the cathode extract. We have found that a large portion of the cations may in some cases collect in the anode extract and anions in the cathode extract, especially at the start of the extraction. Table 5 illustrates this situation with the distribution of K ions over the anode and cathode extracts. During the first five minutes of EUF extraction of the Fjärdingslöv soil almost half of the desorbed K entered the anode extract. In the following fractions approximately one fourth of the desorbed K was found in the anode extract. In the first sub-fraction of the other soils about 10–25% of the desorbed K was found in the anode extract and 5–7% in the following sub-fractions. Since there are differences in this respect between different soils and sub-fractions, we have started to determine the desorbed ions in both the cathode and the anode extracts by pooling them before analysis.

To summarize the results, it was found that the volumes of replicate anode extracts showed good agreement, that the variations among replicate analyses of desorbed P and K were in some cases too high but on the whole acceptable, and finally that the anions and cations were distributed over both cathode and anode extracts especially in the first sub-fractions, which must be taken into consideration when desorbed nutrients are determined.

Acknowledgements   I thank The Foundation for Swedish Plant Nutrition for financial support of this investigation and Mr B. Forsberg for technical assistance.

References

1   Németh K 1972 Bodenuntersuchung mittels Elektro-Ultrafiltration (EUF) mit mehrfach variierter Spanning. Landwirtsch. Forsch. Sonderh. 27, 184–196.
2   Németh K 1976 Die effektive und potentielle Nährstoffverfügbarkeit im Boden und ihre Bestimmung mittels Elektro-Ultrafiltration. Habil Schrift. Giessen.

The effect of P source on P desorption by electro-ultrafiltration (EUF) on two different soils

H. GRIMME and K. NÉMETH

Büntehof Agricultural Research Station Hannover, Federal Republic of Germany

Key words   Electro-ultrafiltration   P desorption   P-source

Summary   P desorption by the EUF method – when carried out with constant field strength – could be described either by zero- or second-order rate equations. In a calcareous soil, different forms of P resulted in different desorption rates. P desorption could be described as a second-order reaction, except when rock phosphate was applied. In an acid soil, P desorption followed exclusively zero-order reaction kinetics and no differences between P forms were observed.

The results of conventional methods did not agree with the EUF data. It was concluded that the EUF method supplies more comprehensive information on the P status of a soil.

Introduction

The present paper deals with P desorption in an external electric field (EUF). P desorption into the soil solution is responsible for the replenishment of the soil solution from which the plant roots absorb their P.

The desorption process is especially important in well buffered soils and for nutrients the concentration of which in the soil solution is rather low, as is generally the case with phosphate.

The normal practice in EUF analysis is to extract the nutrients with 3 different field strengths by increasing the field strength stepwise[3]. The desorption curves thus obtained, however, do not lend themselves to mathematical analysis. Therefore, in this study a constant field strength was employed. The objective was to investigate the effect of P source and soil properties on P desorption and to compare the results with those obtained with conventional soil testing methods.

Material and methods

A manually operated EUF apparatus was employed and the procedure was the same as described for K with a constant field strength of 44.4 V/cm throughout[2]. The soils were also analysed with conventional methods, so that the EUF results could be compared with P contents obtained when extracting soils according to the Ca-lactate (CAL)-, the Double-lactate (DL)- and the H_2O method[4].

Samples of two soils (Table 1) were taken from long-term field plots. These soils had been treated annually for 12 years with 100 kg P_2O_5 in the form of superphosphate, basic slag and rock phosphate[4], so that along with the controls 4 experimental units were included in this study.

The derivation of the equations for a quantitative description of the desorption process has already been presented in another paper[2] and will not be repeated here. In principle, the same equations apply to both K and P desorption. P desorption, however, does not always follow second-order kinetics, as will be discussed in the next section.

Table 1. Some characteristics of the soils employed

	Soil I	Soil II
Soil type	Chernozem (Altoll)	Brown earth (Ochrept)
Parent material	Loess	Granite
pH	7.3	5.5
$CaCO_3 \%$	15	0
Clay %	15	15
Silt %	42	61

Results and discussion

Although the same quantities of P were applied in all treatments, the desorption curves obtained for soil I with the various P sources differed widely (Fig. 1), superphosphate yielding the highest desorbed quantities with basic slag being next. The difference between P_0 and rock phosphate was small.

Except for rock phosphate, P desorption follows second-order reaction kinetics and the desorption curves can be described by the integrated rate equation[1,2]:

$$d = \frac{D \cdot t}{t + t_{0.5}} \tag{1}$$

with:

D = initial content of desorbable P in the soil (mg/100 g)
d = cumulative quantity of P desorbed at time t, extent of reaction (mg/100 g)
t = time (min)
$t_{0.5}$ = half-time, being the time elapsed when 50% of the desorbable P has been removed from the soil

The constants can be computed by using a linear transformation of (1). The curves in Fig. 1 are drawn according to equation (1) and fit the experimental data very closely.

If we now include the results obtained with the conventional methods of analysis in this discussion, we find that neither the CAL nor the DL method differentiates between the superphosphate and basic slag treatments, whereas the H_2O method does (Table 2).

D is equal for superphosphate and basic slag, but the half-time values differ, the one for superphosphate being much shorter. The superphosphate-treated sample releases its P much more readily than does the basic slag-treated sample. This difference is not borne out by the CAL and DL methods, but only by the H_2O method.

The control and rock phosphate-treated samples are rated equal by all

Table 2. Extractable P in soil I as evaluated with different methods

P fertilizer used	P extracted with various methods, mg/100 g soil				Values calculated	
	CAL	DL	H$_2$O	EUF$_{35}$*	D	t$_{0.5}$ (min)
None	5.7	1.7	0.4	1.0	2.2	48
Superphosphate	11.1	6.8	2.0	2.5	3.3	15
Basic slag	11.3	5.6	1.3	1.5	3.3	44
Rock-phosphate	5.8	2.1	0.5	1.2	–	–

* EUF$_{35}$ = total P desorbed in 35 min.

methods except by the kinetic analysis. The control shows a 'normal' desorption behaviour, *i.e.* a desorption rate declining with time (Fig. 3). With rock phosphate, the reaction rate does not decline and the extent of reaction increases linearly with time (Fig. 1). This indicates that the reaction rate is independent of the reacting quantity, *i.e.* desorption takes place from a multilayered P source (such as a rock phosphate particle) the surface coverage of which does not change during the limited time of desorption. It appears that the rock phosphate has remained unaltered in this carbonate-containing soil and dissolves with a constant rate since the dissolution products are continuously being removed.

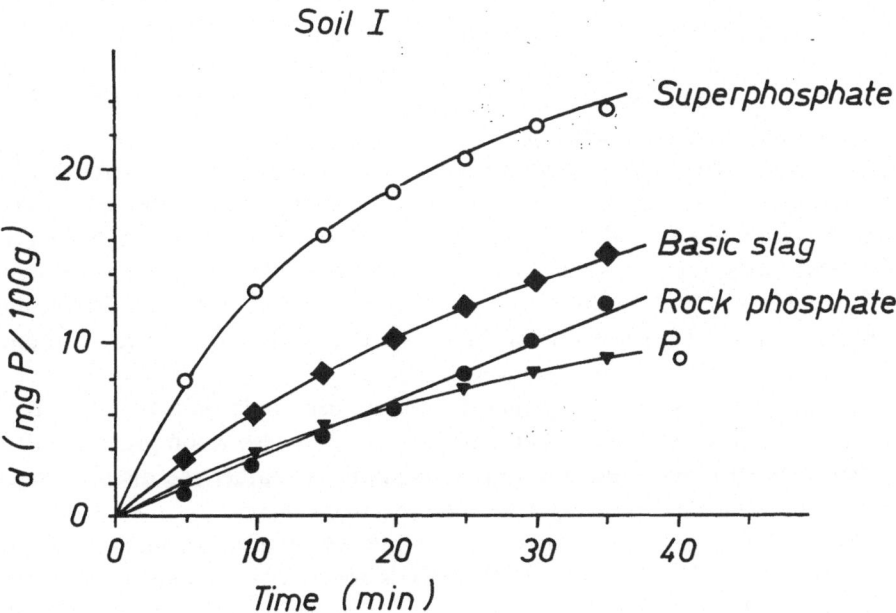

Fig. 1.   Cumulative desorption curves (Soil I).

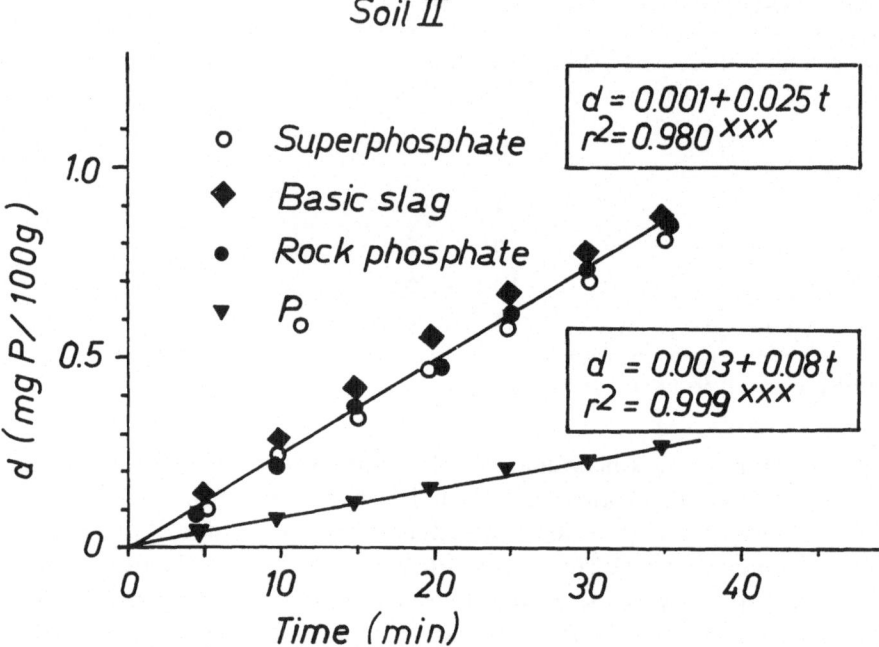

Fig. 2.   Cumulative desorption curves (Soil II).

For soil II, there is practically no difference between the treatments except of course for P_0, which had been depleted of P during the 12 years of cropping. In all cases the desorbed quantity increases linearly with time (Fig. 2), and the desorbed quantities for the superphosphate, basic slag and rock phosphate treatments do not differ significantly so that one desorption curve could be fitted to the data of the three fertilizer treatments.

Apparently the fertilizers have been converted into the same P form in the soil, irrespective of P source. As a consequence, the desorption rates are very similar and they are constant with time. Qualitatively all P sources in this soil behave in the same way as rock phosphate does in soil I which contains $CaCO_3$. In the case of soil I, it was argued that the rock phosphate remained unaltered. The results obtained with soil II show that various P fertilizers can be converted into one form resulting in a constant reaction rate.

The desorption is – within the time interval studied – independent of the extent of reaction. This could mean that the release is not a desorption process with a decrease in surface coverage during desorption, but rather the dissolution of precipitate which has possibly formed on reactive surfaces.

The CAL, DL and H_2O methods extract different quantities from the different P sources (Table 3), whereas the EUF analysis does not differentiate between the sources. Thus, here again a disagreement exists between conventional and EUF analysis. Since none of the three P forms is stable in an acid soil like soil II and all

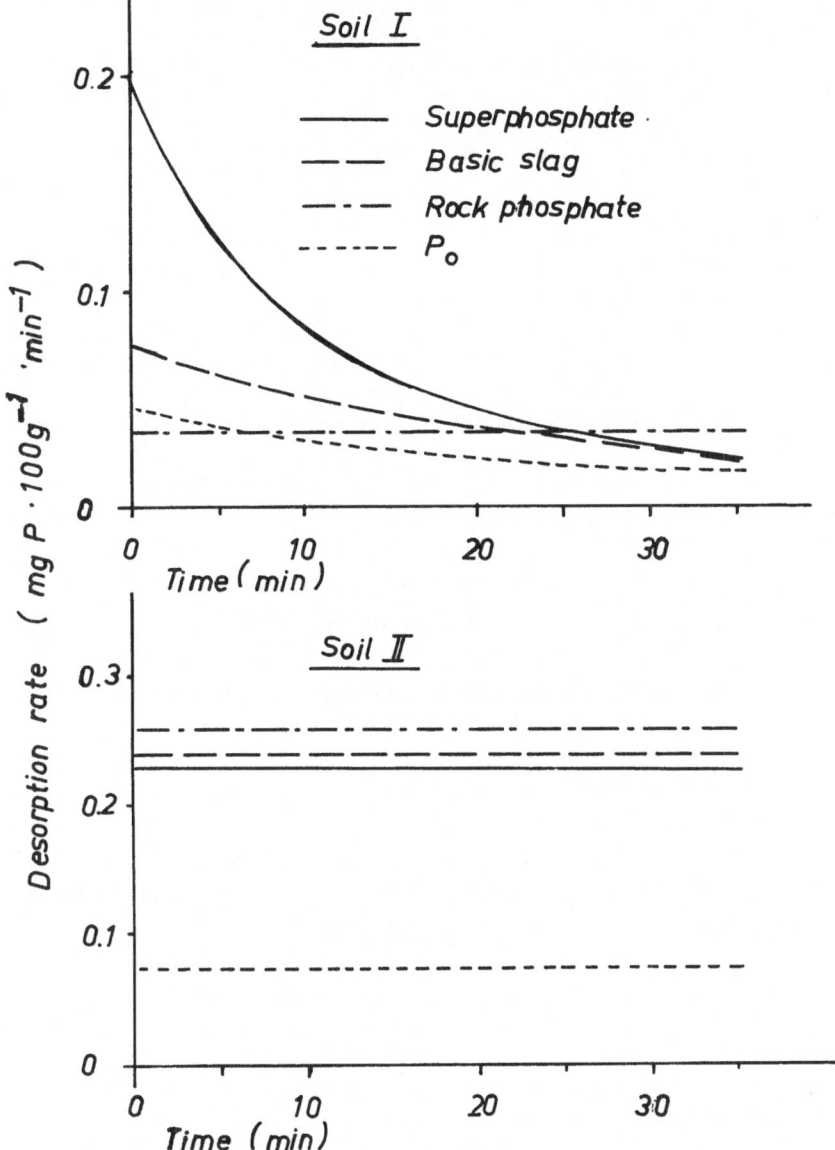

Fig. 3. P desorption rate as a function of time.

three yield the same reaction products, it appears to be quite plausible to observe that the three P forms give similar results.

Because of the constant reaction rate, it is not possible to calculate a maximum desorbable quantity. The soil behaves as a quasi-infinite P source and the P release in soil II proceeds as a zero-order reaction. It was not tested how much P

Table 3. Comparison of P status of soil II as evaluated with different methods

P fertilizer used	P extracted with various methods, mg/100 g soil			
	CAL	DL	H_2O	EUF_{35}*
None	1.4	2.3	0.3	0.27
Superphosphate	3.4	8.2	1.8	0.81
Basic slag	6.6	10.0	2.3	0.88
Rock-phosphate	4.9	12.2	1.6	0.86

* see Table 2

would have to be removed from the soil before a change in reaction rate would occur.

Conclusions

P desorption by the EUF procedure can be described by equations relating the desorbed quantity and the rate of desorption with time of desorption. But unlike that of K, P release does not always follow second-order reaction kinetics. Depending on soil and P form, a zero-order reaction is also possible. This finding can perhaps be used for a study of P forms and transformations in soils.

Since the EUF analysis is in principle a water extraction accelerated by the force of an electric field, the EUF method supplies information on the reactive quantity and its reactivity (reaction rate).

Therefore, the results are often quite different from those obtained with conventional extraction methods. It could be shown in field experiments that the EUF data yielded the closest correlation with P uptake[4].

References

1 Grimme H 1979 The use of rate equations for a quantitative description of K desorption from soils in an external electric field. Z. Pflanzenernaehr. Bodenkd. 142, 57–68.
2 Grimme H 1982 K desorption in an external electric field as related to clay content. Plant and Soil 64, 49–54.
3 Németh K 1979 The availability of nutrients in the soil as determined by electro-ultrafiltration (EUF). Adv. Agron. 31, 155–186.
4 Schüller H, Reichard Th und Németh K 1975 Beziehungen zwischen P-Düngung, Ertrag, P-Entzug und Methoden der Bodenuntersuchung. Landwirtsch. Forsch. 28, 147–157.

K desorption in an external electric field as related to clay content

H. GRIMME

Büntehof Agricultural Research Station Hannover, Federal Republic of Germany

Key words Buffer power Clay content K desorption

Summary The EUF method differs from other soil testing methods in that it not only yields extractable quantities of nutrients but also provides information on the rates with which these quantities are desorbed when an external force is applied. The K desorption can be described by a second-order rate equation. The constants in these equations (maximum desorbable quantity D; halftime $t_{0.5}$) are characteristic for a given K-soil system.

The K desorption rate depends on clay content and on the buffer power of a soil.

Introduction

The EUF method has established itself as a useful tool for the evaluation of soil fertility [7]. It differs from most other methods of soil analysis in that it is a non-equilibrium method. It is the objective of this paper to present equations describing the EUF process in quantitative terms and to relate the constants in these equations to soil properties.

Method and materials

The manually operated EUF apparatus as described by Németh [5,6] was employed. This apparatus is very versatile in that it allows to choose any field strength between 0 and 90 V/cm to extend the desorption period at discretion and to use any time internal between collecting fractions.

Table 1. Some characteristics of the soils used in the investigation

Soil No.	pH	Clay %	Org. C %	Exch. K meq/100 g	K conc.* meq/l
I	6.1	9	1.05	0.84	1.0
II	6.3	34	1.80	0.75	0.2
III	5.8	28	1.95	1.40	1.6
IV	6.7	46	1.90	2.12	0.4
V	6.9	20	1.25	1.04	1.3
VI	3.8	2	7.50	0.15	0.7
VII	6.4	45	5.60	0.11	0.06
VIII	6.6	33	1.65	0.27	0.17
IX	6.5	5	0.78	0.30	0.72

* K concentration in the soil solution.

Desorption time was 50 min, fractions being collected in 5-min intervals. A constant field strength of 44.4 V/cm was applied throughout at a temperature of 20°C. For more details see [1]. The soil sample size was 5 g.

Nine soils were used for this study. Their properties are listed in Table 1.

Results and discussion

K desorption from soils can be described by a second-order rate equation. Since the concentration of the desorbable quantity in the reaction vessel is not known, it is convenient to use the extent of reaction as dependent variable. The equation then reads:

$$\frac{dd}{dt} = k(D - d)^2 \tag{1}$$

in which

D = initial content of desorbable species in the soil ($\mu eq \cdot g^{-1}$)

d = cumulative quantity of ion desorbed at time t, *i.e.* extent of reaction or reaction variable ($\mu eq \cdot g^{-1}$)

t   = time (min)

k = rate constant, a coefficient related to reaction rate.

After integrating and solving for d, the equation reads:

$$d = \frac{D \cdot t}{t + \frac{1}{kD}} \tag{2}$$

The halftime is found by setting d = 0.5 D in (2) yielding:

$$t_{0.5} = \frac{1}{kD} \tag{3}$$

If we substitute $t_{0.5}$ for $\frac{1}{kD}$ we obtain

$$d = \frac{Dt}{t + t_{0.5}} \tag{4}$$

Equation (4) gives the desorbed quantity at time t. It contains 2 constants, D and $t_{0.5}$, which can be assessed by using one of the three linear transformations of (4). The most suitable one was found to be:

$$\frac{t}{d} = \frac{t}{D} + \frac{t_{0.5}}{D} \tag{5}$$

Equation (1) describes the desorption rate as a function of d. The first derivative of (4) then yields an expression for the desorption rate as a function of time.

Table 2. K desorption data obtained at 44.4 V/cm

Soil No.	Exch. K (μeq$\cdot$g^{-1})	D (μeq$\cdot$g^{-1})	$t_{0.5}$ (min)
I	8.4	7.2	7.3
II	7.5	7.2	26.5
III	14.0	13.5	11.4
IV	21.2	22.3	61.7
V	10.4	10.6	12.4
VI	1.5	1.5	2.7
VII	1.1	0.9	40.4
VIII	2.7	2.8	33.1
IX	3.0	3.2	8.4

$$\frac{dd}{dt} = \frac{Dt_{0.5}}{(t + t_{0.5})^2} \tag{6}$$

Desorbed quantity and desorption rate may serve to characterize the ability of a soil to supply a nutrient when it exists in a soil in adsorbed form, such as K and P. Information on the change in desorption rate with time or buffering is obtained from the second derivative.

$$-\frac{d^2d}{dt^2} = \frac{2Dt_{0.5}}{(t + t_{0.5})^3} \tag{7}$$

Thus a set of equations is available for characterizing a soil with respect to the kinetics of nutrient desorption.

The quantity of EUF-desorable K (at 20°C) which is given by D is nearly always identical with exchangeable K[1,2]. The quantity actually desorbed is less than D, since according to equation (4) d = D will only be obtained at t = ∞.

K is preferentially adsorbed by the soil clay[3,4]. The desorption rate, that is the quantity desorbed within a finite time interval will, therefore, depend on clay content and on the K saturation of the clay.

Fig. 1 demonstrates the desorption behaviour of two soils having similar exchangeable K contents but different clay contents (Table 1). The graphs are drawn according to equation (6) using the appropriate constants from Table 2. The soil with a low clay content (I) has a high initial desorption rate which drops rapidly to low values. The high clay soil II has a low initial desorption rate which decreases slowly and overtakes soil I after about 15 min.

On the other hand there are soils having very different exchangeable K contents but nearly the same initial desorption rate (Fig. 2). However, the soil with the lower exchangeable K (and lower D) and lower clay content (IX) is less well buffered, so that its desorption rate drops off very rapidly, whereas the well buffered soil maintains a fairly high rate.

The initial desorption rate and the desorption rate at halftime are a linear

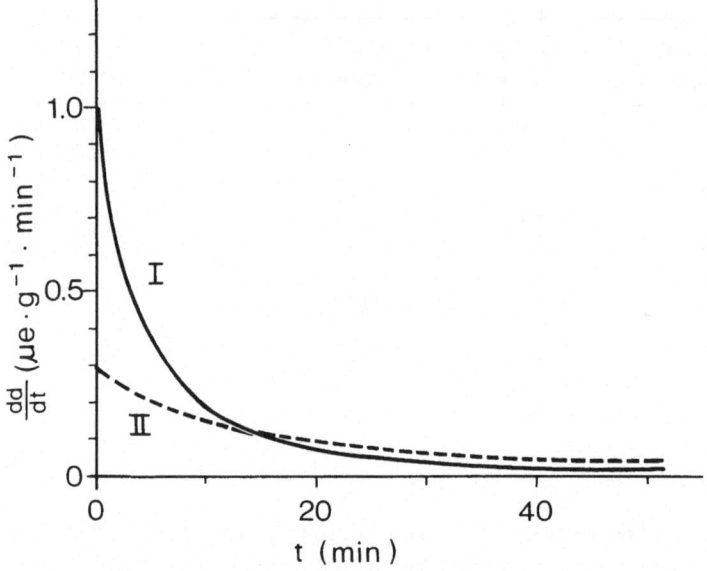

Fig. 1.   K desorption rate as a function of time of two soils with similar exchangeable K but different clay contents.

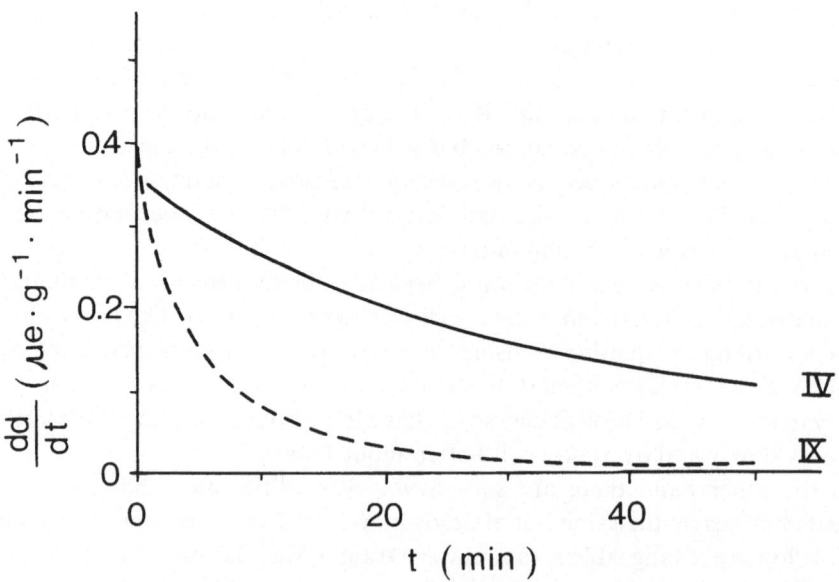

Fig. 2.   K desorption rate as a function of time of two soils with different exchangeable K and clay contents.

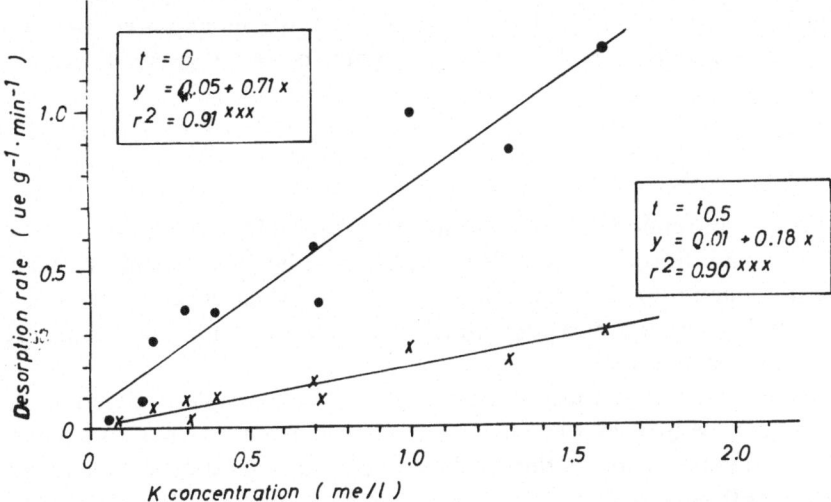

Fig. 3. Relationship between K concentration in the soil solution and initial K desorption rate (t = 0) and K desorption rate at halftime (t = $t_{0.5}$), respectively.

function of the K concentration in the soil solution. The K concentration is a measure of the K saturation of the inorganic exchange sites and of K selectivity (Fig. 3)[3,4].

The halftime is related to the buffer power of a soil, which is characterized by the slope of the adsorption isotherm (Fig. 4). It indicates whether a given quantity

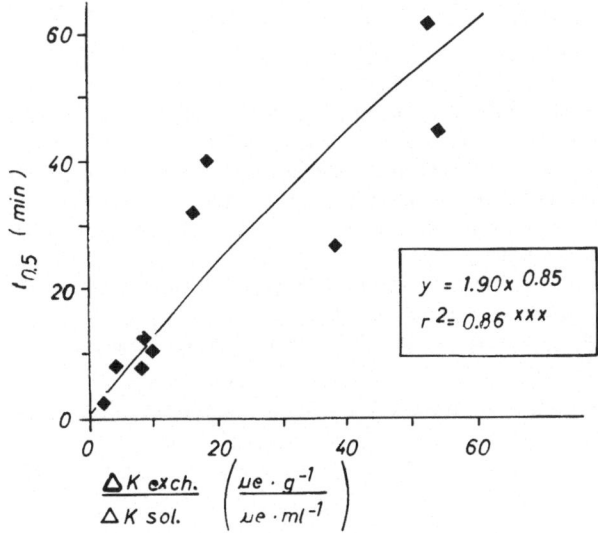

Fig. 4. Relationship between buffer power of soils and halftime.

is released readily or only reluctantly. With respect to K supply it is only meaningful in conjunction with D, in the same way as the buffer power of a soil is not very useful if neither the K concentration nor the exchangeable-K value is known.

Conclusions

Normally, conventional soil analysis provides only information of the extractable quantities of nutrients. With the EUF method it is possible to obtain data on the quantity released within a given time, which means that both a quantity and a rate are measured which can be utilized for a more precise assessment of soil fertility.

When working with a constant field strength it is possible to describe the desorption process by a set of equations that allow the calculation of parameters such as the quantity contributing to desorption and the rate constant or the halftime. It is thus possible to characterize soil K or other nutrients by kinetic parameters. This is of theoretical as well as of practical importance, since the plants do not only require a certain quantity of a nutrient but this quantity must be available within a given time. The desorption process is, of course, only one of several processes being operative in the supply of nutrients to plant roots. Desorption in an electric field may be used to differentiate between quantities of different reactivity and hence mobility.

Literature

1   Grimme H 1979 The use of rate equations for a quantitative description of K desorption from soils in an external electric field. Z. Pflanzenernaehr. Bodenkd. 142, 57–68.
2   Grimme H 1980 The effect of field strength on the quantity of K desorbed from soils by electro-ultrafiltration. Z. Pflanzenernaehr. Bodenkd. 143, 98–106.
3   Németh K, Mengel K and Grimme H 1970 The concentration of K, Ca and Mg in the saturation extract in relation to exchangeable K, Ca and Mg. Soil Sci. 109, 179–185.
4   Németh K and Grimme H 1972 Effect of soil pH on the relationship between K concentration in the saturation extract and K saturation of soils. Soil Sci. 114, 349–354.
5   Németh K 1971 Die Charakterisierung des K-Haushalts von Böden mittels K-Desorptionskurven. Geoderma 5, 99–109.
6   Németh K 1976 Die effektive und potentielle Nährstoffverfügbarkeit im Boden und ihre Bestimmung mittels Elektro-Ultrafiltration. Habil. Schrift, Gießen.
7   Németh K 1979 The availability of nutrients in the soil as determined by electro-ultrafiltration (EUF). Adv. Agron. 31, 155–188.

Effect of soil properties and soil management on the EUF–N fractions in different soils under uniform climatic conditions

T. HARRACH*, K. NÉMETH** and G. WERNER*

Key words EUF–N fractions N_{min} Soil management Soil properties

Summary The influence of soil properties and soil management on the EUF–N fractions in different soils under uniform climatic conditions is described. The results of the investigations are as follows:

1. A close correlation was found to exist between EUF–NO_3 extracted at 20°C and N_{min} contents. These EUF–NO_3 contents were about 0.4–0.8 mg N/100 g higher than the N_{min} contents (Table 2).

2. No correlation was observed between EUF–NO_3 extracted at 80°C and N_{min} contents. These EUF–NO_3 contents, however, showed a certain dependence on the humus and the $N_{Kjeldahl}$ contents (Table 3). They reflect, therefore, soil properties which are not influenced by seasonal changes in weather conditions and by soil management.

3. Application of slurry increases mainly the contents of EUF–NO_3 extracted at 20°C and does not affect the values of EUF–NO_3 extracted at 80°C (Table 2).

4. The EUF-extractable total N fractions (without NO_3) showed a certain correlation with the humus contents as well as with $N_{Kjeldahl}$ values. These total N fractions were smaller in the Anthropic Regosols which are poor in humus than in the other soils rich in humus. The EUF–N fractions (without NO_3) were not influenced by slurry dressings (Table 4).

5. Large amounts of sewage sludge which had been applied to an Orthic Luvisol over a period of ten years considerably increased the EUF–NO_3 (20°C) values as well as the EUF-extractable (20°C) N values (without NO_3). Compared with these values, the increases in $N_{Kjeldahl}$ values were rather small. The EUF–NO_3 (80°C) values were found to be less influenced by additions of sewage sludge than were the EUF–NO_3 (20°C) values (Table 5).

Introduction

In agricultural practice, to make crop production economically profitable a sufficiently high degree of fertilization is necessary. Excessive fertilization, however, may lead to yield depressions or decreases in quality and thus to reductions in profitability of a crop. It may also pollute the environment by leaching of nitrate and other nutrients. It is, therefore, highly important to assess the nitrogen fertilizer amounts to be applied as precisely as possible. Yet, since the conditions governing the microbial processes by which soil organic N is mineralized are very unpredictable, it is rather difficult to calculate the correct amounts of fertilizer to be applied.

Electro-ultrafiltration (EUF) offers a means to fractionate soil nitrogen in such a way that information is obtained on both momentary and potential N

* Institute of Soil Science and Soil Preservation, Justus Liebig University, Gießen, Federal Republic of Germany
** Büntehof Agricultural Research Station, Hannover, Federal Republic of Germany

availability[5]. According to Németh et al.[6], for proper characterisation of the nitrogen status of a soil, the following fractions need to be determined:

EUF–NO$_3$ extractable at 20°C
EUF–NO$_3$ extractable at 80°C
EUF-extractable total N at 20°C
EUF-extractable total N at 80°C

The aim of this paper was to investigate the influence of soil properties and soil management on these individual EUF-extractable nitrogen fractions.

Materials and methods

The investigated soils belong to an area with uniform climatic conditions. This area is situated in the 'Wetterau', a natural landscape unit in Central Germany. The annual average temperature is about 9°C, with a low of 0°C in January and a high of 18°C in July. The annual average rainfall is about 550 mm with a three-month average of about 180 mm from May to July.

The agricultural soils of this area are derived mainly from loess. The following soil units were investigated: Orthic Luvisol (OL), Luvic Phaeozem (LPh), eroded Orthic Luvisol (eOL), strongly eroded loess soil (Calcareous Regosols (CR)). Furthermore Anthropic Regosols (AR) with low contents of organic C were of special interest.

In this area the most important crop from an economic point of view is sugar beet which reaches yields on these fertile loess soils of 48 to 60 t/ha. Sometimes, however, the technological quality of this crop is inferior due to excessive N fertilization. Winter wheat yields (5 to 7 t/ha) are not always satisfactory without an obvious explanation. Some farms without cattle are managed without the use of manure, whereas on other farms with large livestock herds, large quantities of slurry are applied.

Mineral nitrogen (N_{min}) was extracted with 1% KAl(SO$_4$)$_2$ solution[1] and determined by the double destillation method (microdestillation method for fractionated mineral nitrogen analysis) as described by Gerlach[3].

EUF–N fractions were extracted within 0–30 min at 20°C and 200 V (fraction I) as well as within 30–60 min at 80°C and 400 V[4,5] (fraction II).

Results and discussion

The N_{min} contents (NO$_3$ + NH$_4$) of the investigated soils in the summer of 1979 at 5 different dates are presented in Table 1. The high N_{min} values in early summer are due to previous N-additions and to N mineralization. Table 1 clearly shows that the N_{min} contents decrease distinctly from June to September and reach barely measurable values in August/September. These figures are average values, each of them representing 3 replicate samples. The large variability in measured values among soils is rather striking. It may have been caused by the heterogeneity of the soils and of the manure- and fertilizer distribution pattern. In order to obtain more representative results according to Gerlach[3] it would be necessary to analyze ten replicate samples.

In early October 1979, after a rather long dry period, soil samples were taken at 25 sites. The results of analyses are shown in Table 2. The sites are ranked

Table 1. N_{min} contents in the topsoil at 11 sites at different dates

Site no.	Soil* type	Crop** grown	Mineral N extracted, mg/100 g soil				
			24 June	19 July	12 Aug.	3 Sept.	23 Sept.
1	LPh	sb	70.0	79.0	0.6	8.0	0.6
2	AR	sb	36.0	83.0	4.5	0.7	1.8
3	LPh	wb	1.2	1.4			
4	AR	wb	2.5	2.2			
5	AR	al	71.0	2.6	1.2	0.8	1.6
6	AR	al	4.5	1.5	0.0	0.7	0.3
7	AR	al		8.7	1.6	1.1	0.9
8	OL	sb	28.0	5.2	4.8	0.0	0.0
9	OR	sb		2.1	0.8	1.1	1.0
10	OL	sb		2.5	1.6	1.0	1.3
11	OL	sb		11.0	1.5	0.0	0.0

* AR = Anthropic Regosol; CR = Calcareous Regosol; LPh = Luvic Phaeozem; OL = Orthic Luvisol.
** al = alfalfa; sb = sugar beet; wb = winter barley.

according to their N_{min} contents. The values of the first 14 soils are low (0.1–0.5 mg), which is normal for the season[7,8,9]. The subsequent 7 soils show moderate values of N_{min} (0.8–2.8 mg). The higher N_{min} values can be explained partly by high quantities of manure applied. The highest values (3.6–5.7 mg) which were found in the remaining four soils can, however, without any doubt be ascribed to heavy applications of liquid manure or slurry after the wheat crop.

There is no close correlation between the N_{min} contents and other soil characteristics, such as C or $N_{Kjeldahl}$. However, as shown in Table 3, a close correlation was found between N_{min} values and two of the three EUF–NO_3 values.

The significant correlation between the N_{min} and the EUF–NO_3 values at 20°C allows for the conversion of the EUF values into N_{min} values with the aid of the following regression equation:

$$N_{min} = 0.95 \, \text{EUF–}NO_3 \text{ at } 20°C - 0.44$$
$$r = 0.96***$$

The deviations of the N_{min} values, calculated according to the above equation, from the analytically determined N_{min} values are small. In only 4 cases, the deviations were larger than 0.5 mg, and in only one of these cases it was larger than 0.7 mg/100 g soil. A N_{min} content of 0.5 mg corresponds to about 20 kg of nitrogen in a 30 cm layer of soil.

The correlations between the EUF–NO_3 values and the $N_{Kjeldahl}$ and organic C values were not particularly close (Table 3). On the other hand, the correlation between N_{min} and $N_{Kjeldahl}$ values was not significant. It is not surprising that no

Table 2. N fractions and other characteristics of soils at 25 sites

Site no.	Crop** grown	Soil* type	Clay content, %	pH CaCl$_2$	Organ. C %	N$_{Kjeld.}$ mg/100 g	C/N	N$_{min}$ mg/100 g	EUF–NO$_3$ mg N/100 g	
									I	II
1	le	OL	17.1	7.2	1.2	115.4	10.5	0.1	0.8	0.5
2	sb	AR	38.4	7.6	0.8	65.0	12.5	0.1	0.5	0.5
3	al	AR	39.9	7.6	0.4	41.0	10.2	0.1	0.5	0.2
4	le	LPh	30.0	6.6	1.5	155.0	9.7	0.2	1.4	1.0
5	sb	Cr	19.7	7.5	1.4	142.1	9.8	0.2	1.0	0.7
6	le	LPh	23.2	7.4	1.7	163.0	10.4	0.2	1.1	1.0
7	wt	OL	25.1	6.4	1.4	137.9	9.9	0.3	0.8	0.5
8	sb	AR	39.5	7.1	0.7	50.4	13.9	0.3	0.5	0.6
9	wt	LPh	26.1	7.4	1.4	114.7	11.8	0.3	0.9	0.8
10	sb	OL	15.4	5.9	1.3	126.9	10.1	0.3	1.2	0.8
11	sb	LPh	19.2	7.5	1.5	154.3	9.9	0.4	1.2	1.08
12	sb	AR	32.8	7.1	1.1	64.7	16.7	0.5	0.7	1.0
13	sb	CR	25.8	7.1	1.3	151.2	8.5	0.5	1.3	0.8
14	wt	AR	28.8	7.5	1.0	94.5	10.5	0.5	1.2	0.5
15	wt	eOL	26.5	7.0	1.6	165.8	9.8	0.8	1.9	0.8
16	wt	eOL	20.8	7.4	1.5	138.7	10.6	1.0	1.5	0.8
17	sb	AR	36.5	7.6	0.5	47.3	10.9	1.4	1.5	0.7
18	wt	CPh	25.7	7.4	1.4	133.0	10.1	1.5	2.2	0.9
19	sb	eOL	27.8	7.1	1.7	166.5	10.0	2.4	3.3	0.8
20	sb	OL	21.5	7.2	1.3	126.2	9.9	2.6	1.3	1.09
21	sb	LPh	20.6	6.0	1.3	134.5	9.6	2.8	3.1	0.7
22	wts	eOL	28.7	7.5	1.6	156.7	10.3	3.6	4.5	0.8
23	wts	OL	24.7	5.9	1.4	139.4	9.7	4.5	4.7	0.9
24	wts	LPh	33.7	6.1	1.7	151.9	10.9	4.8	5.4	0.9
25	wts	LPh	24.8	6.4	1.8	175.5	10.3	5.8	6.6	0.9

* AR = Anthropic Regosol; CR = Calcareous Regosol; LPh = Luvic Phaeozem; CPh = Calcareous Phaeozem; OL = Orthic Luvisol; eOL = eroded Orthic Luvisol.
** al = alfalfa; sb = sugar beet; wt = winter wheat stubble, tilled; wts = winter wheat stubble, tilled, manured with slurry; le = legumes (intercropping after wheat).

Table 3. Coefficients of correlation between various N fractions in the soils

	N$_{min}$	N$_{Kjeldahl}$	Organ. C	C/N
N$_{min}$	1.00	0.38	0.43*	−0.21
EUF–NO$_3$ at 20°C	0.96***	0.51**	0.53**	−0.24
EUF–NO$_3$ at 80°C	0.21	0.43*	0.45*	−0.13
EUF–NO$_3$ at 20°C and 80°C	0.81***	0.58**	0.53**	−0.25

correlation was found between the C/N values and the EUF–N fractions, since the former values showed relatively little variation within the group of soils investigated.

The data of Table 2 show furthermore that the EUF–NO$_3$ values at 80°C (fraction II) are hardly affected by high amounts of slurry applied (sites nos. 21–25). However, temporary changes in soil management, *e.g.* the application of high amounts of slurry, appear to exert a large influence on the EUF–NO$_3$ contents at 20°C.

A correlation exists between the EUF–NO$_3$ values at 80°C and the organic-carbon values. The investigated 25 soils show the following categories: 5 Anthropic Regosols (C contents not above 1%) having EUF–NO$_3$ (80°C) values ranging from 0.22 to 0.68 mg; 7 eroded loess soils (eOL and CR with C contents between 1.3 and 1.7%) having EUF–NO$_3$ (80°C) values ranging from 0.74 to 0.84 mg, and 9 loess soils (OL and LPh with C contents between 1.3 and 1.8%) having EUF–NO$_3$ (80°C) values ranging from 0.84 to 1.09.

Although the general tendency is rather clear, there are a few exceptions which, however, still show a certain correlation with the humus contents: 3 loess soils (OL and LPh with C contents between 1.2 and 1.4%) having EUF–NO$_3$ (80°C) values ranging only from 0.48 to 0.66 mg and 1 Anthropic Regosol (C content 1.1%) having an EUF–NO$_3$ (80°C) value of 0.98 mg.

Table 4. EUF-extractable N contents in some of the soils listed in Table 3

Site no.	EUF–NO$_3$, mg N/100 g soil		EUF-extractable N (minus EUF–NO$_3$), mg N/100 g soil	
	20°C	80°C	20°C	80°C
2	0.52	0.54	2.13	1.90
3	0.54	0.22	1.79	3.14
5	0.96	0.74	2.24	3.47
6	1.06	0.98	2.46	4.82
8	0.51	0.61	1.34	2.10
9	0.94	0.84	2.47	2.58
10	1.18	0.76	2.69	2.58
11	1.23	1.08	2.24	3.25
12	0.74	0.98	2.02	2.10
13	1.26	0.84	2.24	3.58
14	1.24	0.52	2.46	2.60
18	2.2	0.94	2.13	2.80
19	3.26	0.76	2.46	3.81
20	1.27	1.09	2.24	2.35
22	4.50	0.80	2.02	2.30
23	4.74	0.90	2.20	2.69
24	5.38	0.93	2.24	2.70
25	6.64	0.92	2.58	2.70

As the anodical and cathodical EUF filtrates contain other readily soluble N compounds besides NO_3 and NH_4[6], in some soils the EUF-extractable N contents were also determined. The results are listed in Table 4. For the two Anthropic Regosols poor in humus (C contents 0.4 and 0.7%), the EUF–N are 1.8 and 1.3 mg N, whereas these values ranged from 2.0 to 2.7 mg N in all other soils (C contents 0.8 to 1.8%), in other words these values showed little variation and were not correlated with any of the other soil parameters determined.

The most important finding is that the EUF-extractable N contents (without EUF–NO_3) were not influenced measurably by applications of slurry. This finding applies to the values obtained at 20°C as well as at 80°C. The N compounds in the slurry were obviously quickly oxidized and the resulting NO_3 extracted by EUF at 20°C.

In a ten-year experiment with different levels of sewage sludge, on the contrary, it could be clearly demonstrated that the EUF–NO_3 contents as well as the EUF-extractable N contents (without EUF–NO_3) increased considerably over the years of the experiment (Table 5). It can also be seen from this table that application of sewage sludge to an Orthic Luvisol increases not only the EUF–NO_3 contents but in particular the EUF-extractable (20°C) contents, thus indicating that sewage sludge application also enriches a soil with N compounds that are not immediately oxidized to NO_3. Such an enrichment has a lasting effect on the N supply to crops from the soil pool which is considerably affected by the sewage sludge application. Further investigations on this subject are in progress.

Table 5. EUF–N fractions (mg N/100 g) and Kjeldahl-N values at different levels of sewage sludge over a period of 10 years

Sewage sludge t/ha per year	EUF–NO_3		EUF-extract. N (minus EUF–NO_3)		EUF–N	Kjeldahl-N (mg/100 g)
	20°C	80°C	20°	80°C		
0	0.75	0.54	4.87	5.80	11.96	126
5	2.55	0.65	5.70	5.70	14.30	140
15	5.96	0.66	7.22	7.40	21.24	167
30	7.54	1.02	7.73	7.70	23.99	171

References

1   Bremner J M 1965 Nitrogen availability indexes. In Methods of Soil Analysis. Part II; Agron. 9, 1324–1345.
2   Ellenberg H 1964 Stickstoff als Standortsfaktor. Ber. Deutsch. Bot. Ges. 77, 82–92.

3   Gerlach A 1973 Methodische Untersuchungen zur Bestimmung der Stickstoff Netto-
    Mineralisation. Scr. Geobot. 5, 1–116.
4   Németh K 1976 Die effektive und potentielle Nährstoffverfügbarkeit im Boden und ihre
    Bestimmung mit Elektro-Ultrafiltration (EUF). Habil. Schrift, Fachbereich Angewandte
    Biologie u. Umweltsicherung, Justus Liebig-Universität Gießen.
5   Németh K 1979 The availability of nutrients in the soil as determined by electro-ultrafiltration
    (EUF). Adv. Agron. 31, 155–188.
6   Németh K, Makhdum J Q, Koch K and Beringer H 1979 Determination of categories of soil
    nitrogen by electro-ultrafiltration (EUF). Plant and Soil 53, 407–563.
7   Németh K and Wiklicky L 1980 Erfahrungen mit der EUF-Methode bei der Düngeberatung.
    Kali-Briefe Büntehof 15, 15–35.
8   Scharpf H C 1977 Der Mineralstickstoffgehalt des Bodens als Maßstab für den
    Stickstoffdüngerbedarf. Diss. Fak. f. Gartenbau u. Landeskultur, Techn. Univ. Hannover.
9   Scharpf H C and Wehrmann J 1975 Die Bedeutung des Mineralstickstoffs des Bodens zu
    Vegetationsbeginn für die Messung der N-Düngung zu Winterweizen. Landwirtsch. Forsch.
    Sonderh. 32, 100–114.

Rapid determination of EUF-extractable nitrogen and boron

P. KUTSCHA-LISSBERG and F. PRILLINGER
Tulln Sugar Factory, Tulln, Austria

Key words   EUF–B EUF–NH$_4$ EUF–NO$_3$ EUF-extractable N Rapid determination

Summary   A procedure for the rapid determination of EUF-extractable nitrogen (NH$_4^+$, NO$_3^-$ and easily soluble organic N compounds) is described.

In this procedure the EUF–N fractions are oxidized to NO$_3$. The oxidation with peroxodisulfate is accelerated by ultraviolet (UV) radiation. This reduces the time of digestion to about 15 minutes.

The contents of EUF-extractable N are on the average only between 2–8 mg/100 g soil. Their determination by the new procedure in the form of NO$_3$ is more precise than the results obtained by digestion according to Kjeldahl.

The sum of EUF-extractable N fractions obtained by the new procedure allows to assess the N fertilizer requirements more precisely than is possible when using the EUF–NO$_3$ fractions alone. Therefore this new procedure constitutes a considerable advantage when working out fertilizer recommendations for agricultural practice.

Introduction

The EUF procedure allows for the extraction of NH$_4^+$ and NO$_3^-$ ions and of easily soluble N compounds, such as amino acids, from a soil[3]. Hence the EUF-extractable nitrogen consists of both inorganic and organic N compounds. In routine analysis the determination of the NH$_4^+$ and NO$_3^-$ ions poses no problems. If, however, well aerated soils with a high K supply status are involved, only small amounts of NH$_4$ can be found in the EUF filtrates making a precise determination rather difficult. On the contrary, in soils rich in clay and low in K it is quite easy to measure the contents of EUF–NH$_4$ precisely[2, 3]. This also applies to rice soils[5].

Measurements of soluble organic N compounds in the EUF extracts proved particularly difficult. Digestion of the EUF filtrates in concentrated sulfuric acid (according to Kjeldahl) as employed in common practice is very timeconsuming and hardly suitable for routine analyses as conducted in Tulln, where daily up to 700 nitrogen determinations are made. Furthermore, it has to be taken into account that EUF-extractable N is only between 2–8 mg/100 g soil which involves a considerable risk of erratic results in the back-titration of these quantities. Therefore we used the new method for determining N based on the oxidation of dissolved nitrogen compounds.

The new procedure

As UV radiation accelerates the oxidation of N compounds to NO$_3$ by

Table 1. Recovery of N in different N-containing compounds as measured in the form of NO_3

Compound	N concentration (mg/l)		Recovery
	Theoretical	Observed	
Ammonium-chloride	2.00	2.00	100 %
Glycine	2.00	2.09	104.5%
Asparagine	2.00	2.10	100 %
Glutamine	2.00	2.02	101 %
Alanin	2.00	2.02	101 %
Lysin	2.00	2.00	100 %
Betaine	2.00	1.75	97.5%
Urea	2.00	2.20	100 %
Triethanolamine	2.00	2.00	100 %
Hydroxylammonium-chloride	2.00	2.08	104 %
Sulfanilamide	2.00	1.90	95 %
N–N Dimethylformamide	2.00	2.00	100 %
Tributylamine	2.00	0.90	– %
EDTA	2.00	2.00	100 %
Sodium nitrate	2.00	2.00	100 %
Potassium nitrate	2.00	2.00	100 %
Hydrazinium-sulfate	2.00	0.50	45 %
Potassium-thiocyanate	2.00	1.95	97.5%

peroxodisulfate, it is suggested that this combination be tried for the oxidation of total nitrogen in the EUF filtrates to NO_3. When applied with the use of an Autoanalyzer it was found possible to determine the EUF-extractable N in the form of NO_3 by this new method.

It is known[2] that the EUF filtrates may contain amino acids such as serine, glycine, alanine, asparagine and glutamic acid. Therefore we analysed some synthetic amino acids and other N compounds with this new procedure to examine to what extent these compounds are oxidized to NO_3. The results are presented in Table 1.

This table shows that most of the investigated N compounds can be oxidized for 90–100 per cent to NO_3, thus allowing a determination of their N contents in the form of NO_3. Hydrazinium-sulfate and tributylamine can only be partly oxidized to NO_3, which, however, is of not much importance, as these compounds are not found in soil in measurable concentrations.

Determination of nitrate

Nitrate formed during the oxidation is reduced to nitrite by hydrazine sulfate. Nitrite is finally determined according to the Griess reaction.

The time for digestion is approximately 15 minutes, the subsequent determination of nitrate takes about 10 minutes which results in a total reaction time of 25 minutes.

Determination of boron

The boron concentration in EUF filtrates is usually very low. We used a photometric method for the Autoanalyzer SMA 12/40. A very sensitive agent for the determination of boron is 1.1' dianthrimide which gives a yellowish-green solution in concentrated sulfuric acid. Boron causes the colour to change to blue under anhydrous conditions, the blue colour being used for photometric evaluation. One-tenth of a E.U. theoretically corresponds to 0.08 ppm at 5 cm optical path length which indicates that this method is sensitive enough for boron determination in EUF filtrates. Nitrate, even in very low concentrations, gives a brown colour which can be suppressed by the addition of phenol to the sulfuric acid. Organic matter interferes in a similar way and the method is therefore not applicable to soil samples which contain more than 3 per cent organic matter.

The boron determination with azothine H is not sensitive to any interference by organic matter. Azomethine H is also very sensitive to boron. It forms a yellow ion-associated compound in aqueous solution. The absorbance is read at 420 nm, 0.1 E.U. theoretically corresponds to 0.20 ppm boron (at 3 cm optical path length). The maximum colour development occurs within 1.5–2 hours. Due to our new SMA 6/30 system and a reaction time of 25 minutes, the sensitivity of the method decreases to 0.24 ppm B per 0.1 E.U. which is still sufficient for the determination of boron in EUF filtrates. It is necessary to use a blank to eliminate the influence of unspecified absorbance at 420 nm.

The above-described determination of EUF-extractable nitrogen and boron with the Autoanalyzer 6/30 undoubtedly improved the reliability of the results obtained with the EUF method of soil analysis. The use of a dual sampler will increase the capacity from 30 to 60 (2–30) samples per hour.

Conclusions

Investigations by Németh[2] and Németh and Wiklicky[4] have clearly demonstrated that the EUF-extractable N fractions lend themselves for an assessment of the N status of soils and of N fertilizer requirements more precisely than when the EUF–NO$_3$ fractions are used only. The new procedure for the rapid determination of EUF-extractable N has considerable advantages for making fertilizer recommendations in agricultural practice. Next to determinations of the nutrients Ca, Mg, Na, K, NO$_3$, P, S, B, Mn *etc.* by routine analysis, it is now possible to also measure the sum of the EUF–N fractions rapidly and with precision, so that these results can be taken into account when fertilizer recommendations have to be made.

References

1   Kutscha-Lissberg P and Missbach M 1981 Vollautomatische Bestimmung vom gesamten oxidierbaren Stickstoff in Bodenextrakten. *In preparation.*
2   Németh K 1979 The availability of nutrients in the soil as determined by electro-ultrafiltration (EUF). Adv. Agron. 31, 155–188.
3   Németh K, Makhdum M I, Koch K and Beringer H 1979 Determination of categories of soil nitrogen by electro-ultrafiltration (EUF). Plant and Soil 53, 445–453.
4   Németh K and Wiklicky L 1980 Erfahrungen mit der EUF-Methode bei der Düngerberatung. Kali-Briefe Büntehof 15, 15–35.
5   Wanasuria S, Mengel K and De Datta S K 1980 Preliminary investigations into the relationship between the ammonium extracted by electro-ultrafiltration from wetland rice soil and nitrogen uptake and grain yield of paddy rice. Proc. Int. Symp. on the Application of EUF in the Agricultural Production, Budapest/Hungary.

Fixed NH_4–N in relation to EUF-extractable K

H. W. SCHERER

Institute of Plant Nutrition, Justus Liebig-University, Gießen, West Germany

Key words   EUF extractable potassium   Non-exchangeable ammonium

Summary   Representative arable soils from Hesse were investigated for their contents of fixed NH_4^+ and EUF-extractable potassium in the rooting zone. Alluvial soils were found to be rich in fixed ammonium and low in EUF-extractable potassium, while soils of basaltic origin were low in fixed ammonium and rich in EUF-extractable potassium. A negative correlation ($r = 0.79*$) was found between fixed NH_4^+ and EUF-extractable soil K^+. The content of fixed NH_4^+ in the soil profile showed considerable and significant changes during the growing season, which finding is supposed to be due to NH_4^+ uptake by the crop.

Introduction

Inorganic soil nitrogen is present to a varying extent in the form of ammonium nitrogen of which only a small proportion is easily exchangeable, while the major part is bound rather selectively to 2:1 clay minerals. This latter fraction of ammonium is termed 'fixed ammonium' or 'intercalary ammonium'[4].

Generally the fixation of NH_4^+ is suppressed by the presence of potassium[1, 9], but also the reverse is true in the sense that K^+ fixation is affected by NH_4^+. From these findings it can be deduced that ammonium and potassium are competing for the same selective binding sites[8].

In Germany only few investigations have been carried out on the significance of fixed ammonium, although the quantities of fixed ammonium can be considerable[5, 6].

The aim of our investigation was to examine whether a correlation exists between the contents of fixed ammonium and plant-available potassium. Further, the level of fixed NH_4 was determined throughout the growing season at five different dates.

Material and methods

Relationship between fixed NH_4^+ and exchangeable K^+

Representative arable sites located in Hesse were investigated for their contents of fixed ammonium and exchangeable potassium at soil depths of 0–30, 30–60 and 60–90 cm. Soil samples were taken after harvest. Table 1 supplies information on soil properties and crops involved in this investigation.

Availability of fixed NH_4

Soil samples from a farmer's field under winter wheat were taken at five different dates at the depths

Table 1. Soil characteristics and crops grown

Parent material	Crop	Soil depth (cm)	pH	% clay
Alluvial	Winter wheat	0–30	7.0	32.3
		30–60	7.1	30.2
		60–90	7.5	6.7
Alluvial	Winter wheat	0–30	7.1	32.5
		30–60	7.0	52.5
		60–90	7.1	45.4
Basalt	Winter wheat	0–30	5.8	17.2
		30–60	5.7	17.0
		60–90	5.6	19.1
Basalt	Winter wheat	0–30	6.7	38.1
		30–60	6.4	49.2
		60–90		
Loess	Winter wheat	0–30	6.5	12.4
		30–60	6.0	17.8
		60–90	6.0	30.5
Loess	Maize	0–30	7.1	17.1
		30–60	7.0	19.8
		60–90	7.0	18.1

of 0–30, 30–60, and 60–90 cm throughout the growing season. The sampling dates are listed in Table 2.

Soil analysis

Fixed ammonium was determined according to the method of Silva and Bremner[7], modified by Scherer and Mengel[6].

In the experiment on the availability of fixed NH_4^+, soil samples were also analysed for NH_4^+ by means of EUF at 80°C and 400 V for 35 minutes. For this purpose, the samples were pretreated with KOBr and 0.5 N KCl like described in the method of determination of fixed ammonium.

Potassium was also determined by means of the EUF technique; soil extraction was carried out at 15°C and 3 fractions were collected from which only fraction 1 and 2 were used to characterize the content of nonspecifically-bound potassium.

Results

Table 3 shows the contents of fixed NH_4^+–N and EUF-extractable K^+ in the different soil layers. While the alluvial soils are rich and soils of basaltic origin are

Table 2. Dates of soil sampling and stages of development of winter wheat

March,	3rd	tillering
May,	15th	stem elongation
June,	11th	heading
July,	7th	mature
August,	8th	two days before harvest

Table 3. Contents of fixed NH$_4$–N and EUF-extractable K (1st and 2nd fraction) in the various soils

Parent material	Soil depth	NH$_4$–N ppm (dry soil)	K ppm (dry soil)
Alluvial	0–30	268	42
	30–60	145	10
	60–90	70	13
0–90 cm (kg/ha)		2171	353
Alluvial	0–30	271	32
	30–60	238	75
	60–90	234	24
0–90 cm (kg/ha)		3346	765
Basalt	0–30	56	248
	30–60	32	67
	60–90	25	45
0–90 cm (kg/ha)		524	2100
Basalt	0–30	75	254
	30–60	68	118
	60–90	65	100
0–90 cm (kg/ha)		937	1534
Loess	0–30	116	153
	30–60	130	31
	60–90	137	19
0–90 cm (kg/ha)		1721	1059
Loess	0–30	160	150
	30–60	123	32
	60–90	110	28
0–90 cm (kg/ha)		1763	1144

low in fixed ammonium, the content of EUF-extractable K$^+$ is low in the alluvial soils and rather high in the basaltic soils. The loess soils are ranging intermediately. A clear negative correlation ($r = 79*$) exists between fixed NH$_4$$^+$ and EUF-extractable K$^+$.

Fig. 1 shows the contents of fixed NH$_4$$^+$ throughout a growing period. In the upper two soil layers (0–30 and 30–60 cm) the trend is similar: From the 1st to the 2nd sampling date the fixed (non-exchangeable) NH$_4$$^+$ increased, then a steep

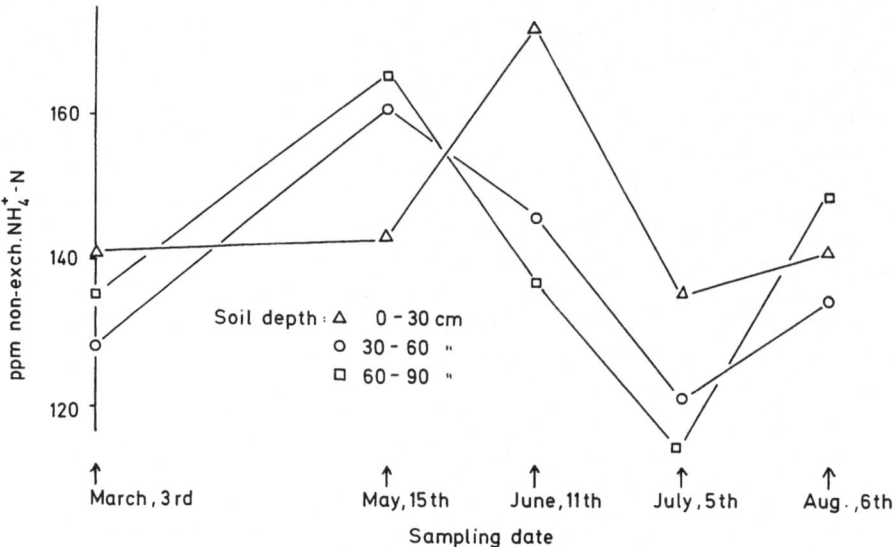

Fig. 1.   Contents of non-exchangeable NH_4^+–N throughout a growing season under winter barley.

decline until July 5th occurred, followed again by an increase in fixed NH_4^+. In the deeper soil layer (60–90 cm), this trend was delayed. Probably plant roots fed from this deeper layer at a later stage. The total loss of fixed NH_4^+ from May to July amounted to about 450 kg N/ha in the rooting zone (0–90 cm).

The fixed (non-exchangeable) NH_4^+ could not be extracted by means of EUF, not even at a temperature of 80°C and 400 V.

Discussion

NH_4^+ fixation is analogous to K^+ fixation. For this reason NH_4^+ and K^+ may compete for the same selective binding sites of the 2:1 clay minerals. According to the investigation of Bartlett and Simpson[2], NH_4^+ application lowers the fixation of K^+.

The aim of this investigation was to find out whether a correlation exists between the contents of fixed NH_4^+ and plant-available K^+. In contrast to other authors, we used the EUF-extractable potassium as an indicator of soil K availability, as the potassium fraction extracted by conventional exchange techniques is not a clear-cut fraction but may contain K^+ from binding sites of different K^+ specificity. According to recent results of Busch[3], the K^+ obtained by EUF extraction (total of 35 minutes) is consisting mainly of K^+ which is nonspecifically adsorbed to clay minerals. This fraction is a more precise indicator of available K^+ as has been shown by Wanasuria et al.[10]

The alluvial soils with their high contents of fixed NH$_4$$^+$ were low in EUF-extractable K$^+$, while the soils of basaltic origin were rich in EUF-extractable K$^+$ but low in fixed NH$_4$$^+$. It is supposed that in the soils of basaltic origin the specific sites of the clay minerals are mainly occupied by K$^+$ and only to a small extent by NH$_4$$^+$. Besides by soil parent material the relationship between the K status of soils and fixed NH$_4$$^+$ may be influenced by fertilizer policy as well. In cases in which K fertilizer application does not match the soil K removed, more binding sites will become available for NH$_4$$^+$ fixation. Such a situation will result in low contents of EUF-extractable K$^+$ and relatively high contents of fixed NH$_4$$^+$.

From Fig. 1 it is apparent, that the fraction of fixed NH$_4$$^+$ was subjected to considerable changes during the growing season. The remarkable decline in fixed NH$_4$$^+$ from May until July is supposed to be due mainly to K$^+$ uptake by the crop. The figure also shows that NH$_4$$^+$ depletion of the clay minerals was partially counterbalanced by NH$_4$$^+$ fixation during a later stage (July until August).

References

1 Atanasiu N, Westphal A and Banerjee A K 1967 Studien über die Wirkung gedüngten Stickstoffs auf Ertrag und N-Aufnahme der Pflanze bei Böden mit verschiedenem Ammoniumfixierungsvermögen. Agrochimica 11, 257–274.

2 Bartlett R J and Simpson T J 1967 Interaction of ammonium and potassium in potassium-fixing soil. Soil Sci. Soc. Am. Proc. 31, 219–222.

3 Busch R 1980 Der Einfluß der K$^+$-Konzentration der Bodenlösung und der K$^+$-Pufferung auf die K$^+$-Aufnahme und das Wachstum von Lolium multiflorum. Diss. Gießen.

4 Osborne G J 1976 The extraction and definition of nonexchangeable or fixed ammonium in some soils from southern New South Wales. Aust. J. Soil Res. 14, 373–380.

5 Schachtschabel P 1961 Bestimmung des fixierten Ammoniums im Boden. Z. Pflanzenernaehr. Dueng. Bodenkd. 93, 125–136.

6 Scherer H W and Mengel K 1979 Der Gehalt an fixiertem Ammoniumstickstoff auf einigen repräsentativen hessischen Standorten. Landwirtsch. Forsch. 32, 416–424.

7 Silva J A and Bremner J M 1966 Determination and isotope-ratio analysis of different forms of nitrogen in soils. Soil Sci. Soc. Am. Proc. 30, 587–594.

8 Sippola J, Erviö R and Eleveld R 1973 The effect of simultaneous addition of ammonium and potassium on their fixation in some Finnish soils. Ann. Agric. Fenn. 12, 185–189.

9 Stanford G and Pierre W H 1946 The relation of potassium fixation to ammonium fixation. Soil Sci. Soc. Am. Proc. 11, 155.

10 Wanasuria S De Datta S K and Mengel K 1980 Rice yield in relation to electro-ultrafiltration extractable soil potassium. Plant and Soil 59, 23–31.

Preliminary investigations on available and potentially available phosphorus using electro-ultrafiltration (EUF)*

M. A. TAHA, M. N. MALIK, M. I. MAKHDUM and F. I. CHAUDHRY
Physiology/Chemistry Division, Central Cotton Research Institute, Multan, Pakistan

Key words Buffering capacity EUF extracted phosphorus Readily available phosphorus Reserve phosphorus Seed cotton yield Slowly available phosphorus

Summary Preliminary investigations were carried out on calcareous silty loam (clay content 25%, pH 8.0) at the Cotton Research Institute, Multan to determine the solubility and desorption rates of phosphorus at various soil depths throughout the cotton growing season using the EUF technique. The treatments included two applications of single superphosphate, equivalent to 0, 75, 150, 250 and 350 kg P_2O_5/ha. There was no significant difference in seed cotton yield between the five treatments.

EUF-extracted phosphorus decreased with depth in all samples. There was an increase in phosphorus concentration during the growing season, but most of the increase was noted in the slowly available phosphorus fraction (10–30 minutes EUF aliquots). This was attributed to continuous transformation of phosphorus into less soluble calcium phosphate forms. The available phosphorus fraction (0–10 minute EUF aliquots) remained small throughout the growing season. It was concluded that this fraction was in equilibrium with the slowly available phosphorus fraction and was continually being replenished as crop uptake continued. A high level of effectively available phosphorus (0–30 minute EUF aliquots) was recovered in all five treatments and was above the sufficiency level of 12 ppm phosphorus reported by Nemeth and Makhdam[7]. There was therefore no difference between treatments in phosporus concentration nor in phosphorus uptake by plants.

Introduction

Comprehensive reviews have been published on the phosphorus fertilization of cotton by Wahab[12], Ali[1], and Khan[6]. Each concluded that there was little or no response to the addition of phosphoric fertilizers to cotton in Pakistan. Nevertheless, the recommended practice for the cotton area stands at 40 to 50 lbs P_2O_5/acre. Nearly all phosphate fertilizer used in Pakistan is imported from abroad, thus substantial savings in foreign currency could accrue if recommendations could be based on the results of scientific research. The reasons for the lack of response to phosphorus in the soils of Pakistan are however not understood. It was therefore decided to investigate in more detail the underlying reasons for this lack of response by the cotton plant.

The standard method of phosphate extraction in Pakistan is Olsen's method[4] which uses 0.5 M $NaHCO_3$ as an extracting solution. Because the extracting solution is alkaline (pH 8.5) and has relatively little effect on other less available calcium phosphates, the method is particularly suitable for use with calcareous soils such as those obtained in Pakistan. $NaHCO_3$-extractable phosphorus

* This work was carried out under UNDP/FAO Project-PAK 73/026

varies from 5–25 ppm in different cotton soils of Pakistan[3]. Chaudhry[3] stated that the limits established for soils which will respond to the application of phosphorus or will probably respond are 0–4 and 4–8 ppm, respectively. Khan[5] did not find any response to phosphorus in soils varying from 3.0 to 19.5 ppm $NaHCO_3$-extractable phosphorus. He concluded that a range of 3.0 to 6.0 ppm $NaHCO_3$-extractable phosphorus was sufficient to meet the demands of the cotton crop. Németh and Makhdum[7] determined the effectively available phosphorus of different soils from Pakistan extracted by the Electro-Ultrafiltration apparatus (EUF) at 20°C and 200 volts. They concluded that these soils contained medium to good phosphorus supplies and postulated that for optimal plant nutrition a level of 12 ppm EUF-phosphorus extracted at 20°C and 200 V was sufficient.

Phosphorus availability to crop plants depends on soil properties which determine its solubility and desorption. These include soil texture (mainly clay content), soil reaction (pH), and $CaCO_3$, and a separate determination of all these is important. Most of the conventional methods employed to extract the available phosphorus in soils do not take these properties into consideration. A new technique based on the use of an electric field to separate nutrient fractions in a soil suspension has recently been developed[8]. The advantage of the Electro-Ultrafiltration technique (EUF) over the conventional methods is that separate determinations of other soil properties are not required, and that from a single extraction, the desorption- and solubilization rates of several nutrients important to plant growth can be determined. These desorption rates give an indication of effective and potential availability of nutrients[8]. Such an apparatus has been installed in the physiology/chemistry laboratory of the Central Cotton Research Institute, Multan. A preliminary investigation to determine the solubility and desorption rates of phosphorus at several soil depths and at various times throughout the cotton growing season was conducted.

Materials and methods

The cotton variety B 557 (*Gossypium hirsutum* L.) was sown in four replicate 3.65 × 12.95 m plots arranged in a complete randomised block design. Each treatment received the equivalent of 100 kg N/ha as urea applied at sowing. The soils were an alluvial silty loam with a clay content of 25% and a pH 8 (soil : water, 1 : 10). Single superphosphate at the rates of 0, 50, 100, 150, and 200 kg P_2O_5/ha was applied on the second of July immediately before sowing. The heavy monsoon rainfall during July, however, completely washed out the young crop and the whole experiment had to be resown on the first of August. Second single superphosphate applications equivalent to 0, 25, 50, 100 and 150 kg P_2O_5/ha were superimposed on the first applications and were broadcast before the first irrigation 30 days after resowing.

To estimate the initial soil phosphorus content, five soil cores were removed from the experimental area before the first sowing and before the application of treatments. Soil samples of the 0–30, 30–60 and 60–120 cm depths were collected from the cores. At approximately one-month intervals, subsequent samples were taken from two replicates. These samples consisted of a composite of three cores from each plot, removed depthwise as in the initial sampling. The soil from each stratum was thoroughly mixed, air dried and passed through a 2-mm sieve. Five-gram samples were taken from each sample for analysis.

The samples were extracted by EUF with stepwise increase in voltage. Seven EUF aliquots were collected at 5-minute intervals. The first aliquot (0–5 min) was extracted at 50 volts, the next five aliquots (5–30 min) were extracted at 200 volts and a single 5-minute aliquot (30–35 min) was extracted at 400 volts. The extracted aliquots were analysed using the chlorostannous molybdophosphoric blue colour method in a hydrochloric acid system. The absorbance was measured in 10-mm cells at 660 nm. on the spectrophotometer.

Four plants, cut at ground level, from each plot were removed on five sampling occasions during the growing season, corresponding in time to the soil sampling. Plant material was dried to constant weight at 90°C in a forced-draft oven, and was finely ground. Total phosphorus concentration in plant tissue was estimated using wet digestion and colour development according to the molybdate-vanadate method[11]. The absorbance was read on a spectro-photometer at 420 nm.

Results and discussion

The phosphorus quantities obtained in the first two EUF aliquots (0–10 min) were summed and considered to represent the readily available phosphorus fraction (RAP), with the combined quantities in the four remaining aliquots (10–30 min) representing the slowly available phosphorus fraction (SAP). The phosphorus in the last aliquot (30–35 min) was considered to represent the reserve phosphorus fraction[9] (RP).

Mean values of these fractions obtained for the initial sample before treatment application are presented in Table 1.

It will be seen from Table 1 that the RAP increased with soil depth and the phosphorus in the other two fractions (SAP and RP) decreased with soil depth. The reduced RAP in the top layer is likely to reflect a lower moisture content of this dry layer in comparison to the deeper layers which contained more moisture at this time of the year, and therefore may have released their P somewhat more readily. The increased value in the 10–30 minute extract in the top layer may be taken as evidence of more P retention as the soil dries out.

There was no obvious effect of treatment on the size of the EUF-P fractions obtained from the five samples taken at different stages of growth. The values of these EUF fractions were therefore pooled to give estimates of P concentration averaged over the growing season (Table 2).

Table 1. Phosphorus status of the soil used in the experiment before sowing and treatment application (ppm)

Soil depth cm	Readily available phosphorus	Slowly available phosphorus	Reserve phosphorus	Total
0–30	1.97	7.18	2.41	11.56
30–60	2.75	4.19	2.41	9.91
60–120	3.81	3.94	1.94	6.69

Table 2. Soil P status at different depths during the cotton growing season (averages of 5 dates of sampling)

Treatment Kg P_2O_5/ha	Soil depth cm	EUF-Extracted phosphorus (ppm)			
		Readily available	Slowly available	Reserve	Total
0	0–30	3.82	14.82	3.91	22.34
	30– 60	3.71	8.86	3.60	16.17
	60–120	2.63	4.92	1.34	8.89
75	0– 30	3.70	14.61	5.81	24.12
(50 + 25)	30– 60	3.85	11.52	3.66	19.03
	60–120	3.69	9.65	3.08	16.42
150	0– 30	2.92	14.32	3.82	21.06
(100 + 50)	30– 60	3.00	10.98	3.77	17.75
	60–120	2.95	6.19	2.66	11.80
250	0– 30	4.29	18.11	4.45	26.85
(150 + 100)	30– 60	4.16	12.41	3.16	19.73
	60–120	2.86	7.04	3.14	13.04
350	0– 30	4.13	12.87	4.30	21.30
(200 + 150)	30– 60	2.96	7.22	2.38	12.56
	60–120	3.99	5.95	2.49	11.40

During the growing season the depthwise decrease in EUF-extracted phosphorus was more obvious than in the pre-sowing sample. Table 2 also shows that the size of each phosphorus fraction was much larger than in the pre-sowing sample. The 0–10 minute extract (RAP) showed the least change but larger increases were noted in the 10–30 minute fraction. Németh[9] suggested that high values in this fraction indicated good buffering capacity of the soil and that the sum of the 0–30 minute fractions (RAP + SAP) may be taken as a measure of the effectively available phosphorus. These were equally high in all treatments irrespective of the amount of single superphosphate added.

The discrepancy between the amounts desorbed in the pre-sowing sample and the post-sowing samples indicates that in the dry state and when there was no crop only limited quantities of phosphorus were solubilized, but with frequent irrigation, crop uptake and increased root activity solubilization was greatly enhanced (cf control treatment Table 2). It further indicates that soil contains larger amounts of phosphates than can be extracted by the EUF at the temperature and voltage employed (room temperature and 200 volts for most of the extraction time). Németh and Makhdum[7] working with soils from Pakistan recovered more EUF-phosphorus when the extraction was carried out at 80°C and 400 volts than at 20°C and 200 volts. They concluded that the high values extracted at 80°C and 400 volts constituted soil phosphorus reserves and to a large extent determined the change in the effectively available phosphorus content of these soils.

In plots treated with single superphosphate rapid fixation must have taken place. Sharif et al.[10] concluded that added superphosphates in these soils were rapidly transformed into calcium phosphate products approaching in their solubility that of octocalcium phosphate and that change into less soluble phosphorus forms (apatite) was very slow. It is therefore to be expected that as the season progresses differences between treated and untreated plots in extracted P will diminish as can be seen from Table 2.

Values on phosphorus concentration, expressed as percentage on a dry weight basis, and on calculated P uptake at different stages of growth are shown in Tables 3 and 4, respectively. There was no effect of treatment on both values. Crop uptake of P increased as the season advanced because of the progressive increase in total biomass.

There were likewise no significant differences in seed cotton yield between treatments (Table 5). A reduced yield associated with high P application was previously reported in Multan soils[1,2].

These results confirm the lack of response to P reported frequently in Multan soils. It is obvious from the EUF results presented in Table 2 that all treatments

Table 3. Effects of P application on phosphorus concentration in plant material

Kg P_2O_5/ha	P contents of plant material, % of dry matter on different days after sowing					
	21	47	79	110	139	174
0	0.37	0.32	0.35	0.35	0.31	0.36
75	0.37	0.33	0.38	0.42	0.33	0.38
150	0.38	0.35	0.35	0.35	0.36	0.35
250	0.38	0.35	0.39	0.35	0.37	0.30
350	0.38	0.38	0.40	0.36	0.36	0.30

Table 4. Effect of P application on phosphorus uptake at different stages of growth

Kg P_2O_5/ha	Yield of P (kg/ha) on different days after sowing					
	21	47	79	110	139	174
0	0.10	1.28	17.93	24.45	30.41	32.94
75	0.10	1.57	24.17	28.81	35.51	40.09
150	0.11	1.77	19.69	27.69	40.94	37.50
250	0.10	1.71	18.26	30.29	42.68	31.71
350	0.10	1.96	19.32	23.89	34.20	32.11

Table 5. Effect of P application on seed cotton yield, kg/ha

Treatment kg/ha P_2O_5	Yield kg/ha
0	863 (9.35)*
75	901 (9.77)
150	968 (10.11)
250	950 (10.30)
350	854 (9.26)

*(Yield in mounds per acre in parentheses)

including the control indicate good buffering capacities as measured by the high values of the 10–30 minute EUF fractions. This good buffering capacity keeps the soil solution sufficiently replenished with RAP throughout the season.

References

1   Ali M 1970 Some thoughts on the fertilizer practices in cotton. Ziraat Nama 9 (17) in Urdu (English translation CRI, Multan).
2   Ali M 1976 Effect of application of phosphatic fertilizers on the yield of cotton at Multan and substations. Pak. Cottons 20, 235–243.
3   Chaudhry T M 1972 Cotton soils of Pakistan In Cotton in Pakistan. Pakistan Central Cotton Committee, Karachi, Pakistan 287–294.
4   Jackson M L 1962 Soil Chemical Analysis. Constable & Co. Ltd., London, 1962, 162–165.
5   Khan M A 1975 Studies on the availability of phosphorus in soil and its uptake by the cotton plant at different stages of growth. Fifth Ann. Rep. Physiology/Chemistry Division, Central Cotton Research Institute, Multan, Pakistan, 223–225.
6   Khan M A 1979 Phosphorus nutrition of the cotton plant. Pak. Cottons 23, 147–160.
7   Németh K and Makhdum M I 1981 Evaluation of the nutrient dynamics in calcareous soils from Pakistan by Electro-Ultrafiltration (EUF). Soil Sci. Plant Nutr. 27(2), 159–168.
8   Németh K 1976 The effective and potential availability of nutrients in soil and their determination by Electro-Ultrafiltration (EUF). Appl. Sci. Developm. 8, 89–111.
9   Németh K 1977 The determination of desorption rates of nutrients in the soil by means of Electro-Ultrafiltration (EUF) and its importance for crop production. Proc. Int. Seminar on Soil Environment and Fertility Management in Intensive Agriculture, Tokyo, Japan 1977, 803–810.
10 Sharif M, Chaudhry F M and Lakho A G 1974 Suppression of superphosphate phosphorus fixation by farmyard manure II. Some studies on the mechanisms. Soil Sci. Plant Nutr. 20, 395–401.
11 Shouichi Y, Douglas A F, James H C and Kwanchai A G 1972 Laboratory Manual for Physiological Studies of Rice. The International Rice Research Institute, Manila, Philippines, 13–14.
12 Wahab A 1960 Soil, fertilizer and microbiology. Fifty years of agricultural education and research at Punjab Agricultural College and Research Institute, Lyallpur. Agric. Univ., Faisalabad, Pakistan.

Influence of soil type and potassium and magnesium fertilization on the release of potassium and magnesium by electro-ultrafiltration and ammonium acetate–lactate extraction

MAGNUS HAHLIN

Dept. of Soil Sciences, Swedish University of Agricultural Sciences, S-750 07 Uppsala 7, Sweden

Key words Ammonium acetate-lactate (AL) method Electro-ultrafiltration Magnesium Potassium Soil analysis

Summary The effect of increasing potassium and magnesium fertilization during 15 years on the amounts of K and Mg extracted by AL-solution and desorbed by Electro-Ultrafiltration (EUF) was determined in four Swedish soils.

In all soils, with increasing potassium fertilization the amounts of K extracted by AL and desorbed by EUF increased, and the amounts of Mg–AL and Mg–EUF decreased. Magnesium fertilization had no significant effect on the K values but increased both Mg–AL and Mg–EUF.

The EUF-values were lower than the corresponding AL-values, but the ratio of EUF-desorbed to AL-extracted amounts of K and Mg varied depending on soil type as well as on fertilization rate. The ratio of K–EUF to K–AL increased and the ratio of Mg–EUF to Mg–AL decreased with increasing potassium fertilization, whereas magnesium fertilization decreased the ratio of Mg–EUF to Mg–AL.

Introduction

In Sweden a soil testing system for agricultural soils has been established whereby the lime, phosphorus and potassium status are always determined, but the contents of magnesium, boron, copper and manganese can also be analysed. Potassium, magnesium and phosphorus are determined by extracting the soil with a well buffered, moderately acidic (pH = 3.75) ammonium acetate-lactate solution (AL-solution)[1]. According to unpublished data the AL-extraction gives fractions of K and Mg which roughly correspond to what will be referred to as the exchangeable K and Mg, respectively.

The data from soil analysis have been evaluated in a large number of field experiments carried out on different soil types all over the country. The results of these experiments have shown that the amount of AL-extractable K in topsoil usually gives a satisfactory description of the potassium status and hence an acceptable base for K-fertilizer recommendations[2]. Similarly, the amount of AL-extractable Mg gives a satisfactory base for Mg-fertilizer recommendations. However, the correlation between soil analysis data and fertilization effects is very low[3]. Some of the correlation discrepancies may be ascribed to an unbalanced supply of potassium and magnesium[4]. A large number of irregularities may, however, be caused by the fact that the AL-method, like most soil analysis methods used today, determines a definite fraction of nutrient based on chemical equilibration that reveals nothing about the availability of nutrients to plants. Nor can the method describe the relative ability of the soil to maintain a determined level of nutrients.

The use of the Electro-Ultrafiltration method (EUF) according to Németh[5, 6] seems, however, to provide the possibility of determining both the immediately available fraction as well as the rate at which more or less fixed and exchangeable potentially available nutrients will be transformed into plant-available form.

The present paper gives some results of experiments on the influence of potassium and magnesium fertilization on the amounts of EUF-desorbed K and Mg compared to the AL-extracted amounts.

Materials and methods

Soil samples in this investigation were collected in the autumn of 1978 from differently treated plots of four long-term field experiments. The experiments were started in 1964 and the plots received an annual treatment with 0, 40, 80 or 160 kg per ha. In 1975 two of four replicates were treated with 80 kg Mg per ha. Some basic data on the experimental soils are given in Table 1. The extraction and analyses of the AL-solutions were made by the Swedish National Laboratory for Agricultural Chemistry, whilst EUF-extractions and analyses were performed at the Department of Soil Sciences.

An automatic EUF equipment, model 723, was used for extracting the soil samples. The mass of the samples was 5 g and an extraction time of 35 minutes was used, but extracts were collected and analysed separately every 5 minutes. The voltage used was altered during the extraction time and was 50 V for the first 5 minutes, then 200 V from 5 to 30 minutes and 400 V from 30 to 35 minutes. The volume of the cathode extract plus rinse water was approximately 50 ml per 5-minute extraction time. The concentrations of K and Mg in the cathode extract were determined on a flame emission spectrometer and by atomic absorption, respectively.

Results and discussion

Table 2 shows the amounts of potassium and magnesium extracted by the AL-solution as well as desorbed by EUF during 35 minutes. In all soils there is a

Table 1. Some basic data of the experimental soils. The figures relate to analyses at the start of the field experiments in 1964

Soil	Mechanical composition, %				Chemical properties			
	Humus	Clay	Silt	Sand	pH(H$_2$O)	P–AL	K–AL	Mg–AL
						mg/100 g		
C 10	5	48	26	21	6.7	3.9	21.0	42.0
R 149	4	40	27	29	6.2	8.5	24.0	17.2
R 154	4	7	53	36	5.6	2.3	8.2	8.0
L 305	2	2	42	54	5.6	7.6	9.0	2.5

Table 2. Amounts of AL- and EUF-extractable K and Mg, in mg per 100 g of air-dry soil, in topsoil of four Swedish field trials

Soil	Treatment, kg/ha		Extracted by AL			Desorbed by EUF			Desorbed by EUF % of AL-extractable	
	K Yearly 1964–78	Mg 1975	K	Mg	K/Mg quotient	K	Mg	K/Mg quotient	K	Mg
C 10	0	0	19.6	46.	0.4	6.2	4.46	1.4	31.7	9.7
	40	0	19.5	39.	0.5	6.0	3.92	1.5	30.8	10.1
	80	0	22.5	38.	0.6	6.5	3.82	1.7	28.9	10.1
	160	0	26.5	39.	0.7	8.4	3.50	2.4	31.7	9.0
	0	80	19.5	45.0	0.4	5.9	3.84	1.5	30.2	8.5
	40	80	20.5	40.	0.5	6.2	3.40	1.8	30.2	8.5
	80	80	22.0	37.	0.6	6.2	3.49	1.8	28.2	9.4
	160	80	25.0	40.	0.7	8.9	3.34	2.7	35.6	8.3
R 149	0	0	21.5	16.1	1.3	9.2	2.36	3.9	42.7	14.7
	40	0	25.0	15.3	1.6	11.4	2.13	5.4	45.7	13.9
	80	0	31.0	15.4	2.0	17.4	2.18	8.0	46.2	14.2
	160	0	39.0	12.5	3.1	17.8	1.84	9.7	45.7	14.7
	0	80	25.0	16.8	1.5	10.0	2.44	4.1	40.0	14.5
	40	80	27.0	15.6	1.7	12.6	2.26	5.6	46.7	14.5
	80	80	26.0	18.2	1.4	12.0	2.57	4.7	46.1	14.1
	160	80	37.0	16.2	2.3	18.4	2.24	8.2	49.8	13.8
R 154	0	0	8.0	4.7	1.7	5.0	1.18	4.2	62.5	25.1
	40	0	12.0	5.5	2.2	8.7	1.34	6.5	72.5	24.4
	80	0	18.5	3.5	5.3	12.2	1.06	11.5	65.8	30.3
	160	0	30.5	3.5	8.7	18.5	0.71	26.1	60.6	20.3
	0	80	12.5	7.9	1.6	7.0	1.92	3.6	55.9	24.3
	40	80	18.0	6.4	2.8	11.7	1.45	8.1	64.9	22.7
	80	80	22.5	6.6	3.4	13.6	1.57	8.7	60.6	23.8
	160	80	34.5	7.2	4.8	24.1	1.54	15.6	69.9	21.4
L 305	0	0	4.0	0.9	4.4	2.9	0.43	6.7	72.5	47.8
	40	0	6.5	0.7	9.3	4.2	0.36	11.7	64.5	51.5
	80	0	9.0	0.8	11.2	5.8	0.42	13.8	64.5	52.6
	160	0	15.0	1.1	13.6	9.2	0.38	24.2	61.3	34.6
	0	80	3.5	2.0	1.8	2.3	1.06	2.2	65.8	52.9
	40	80	6.5	2.0	3.2	3.9	0.83	4.7	59.8	41.5
	80	80	8.5	1.8	4.7	5.6	0.91	6.2	65.8	50.5
	160	80	13.5	1.8	7.5	9.0	0.78	11.5	66.7	43.3

significant relation between potassium fertilization and the extracted amounts of both K and Mg. The amounts of K–AL as well as K–EUF increased with increasing potassium fertilization, while the amounts of Mg–AL and Mg–EUF decreased. On the other hand, magnesium fertilization increased both Mg–AL and Mg–EUF, but had no significant effect on the K-values.

The effect of fertilization on the extracted amounts of K and Mg varied between soils. The smallest effects on the extracted amounts of K and Mg were obtained in rather heavy clay soils (C 10 and R 149) with high contents of both K and Mg, while the effects were more pronounced in both a loamy silt (R 154) and a silty sand (L 305) poor in both potassium and magnesium.

In all soils, the amounts of potassium and magnesium desorbed with EUF were less than the amounts extracted in AL-solution, but the ratio of EUF-desorbed to AL-extracted amounts of K and Mg varied and depended not only on soil type but also on fertilization. In the very fertile clay soil (C 10) only about 30 per cent of the AL-extractable K and about 10 per cent of Mg–AL were desorbed by EUF. In the other clay soil (R 149), about 45 and 15 per cent of exchangeable K and Mg, respectively, were desorbed. In the silty and sandy soils, up to about 70 and 50 per cent of AL-extractable K and Mg, respectively, were desorbed.

In all soils potassium fertilization had increased the ratio of EUF-desorbed to AL-extracted potassium and had decreased this ratio for magnesium. In contrast, magnesium fertilization seems to have the ability to decrease the ratio for Mg, while its effect on K-desorption is inconsistent.

The ratios of K and Mg desorbed by EUF-extraction differed in all soils from the corresponding ratios in the AL-extract. The K/Mg-quotient of the EUF-extract was, on average, two to four times that of the AL-extract. Obviously, EUF desorbs exchangeable K more easily than exchangeable Mg. In fact, the ratio of

Table 3. Potassium and magnesium desorption by EUF and extraction by AL-solution in some soils from plots treated yearly with 80 kg K per ha for 15 years. Release of K and Mg in mg per 100 g air-dry soil

Extract	Soil C 10			Soil R 149			Soil R 154			Soil L 305		
	K	Mg	K/Mg	K	Mg	K/Mg	K	Mg	K/Mg	K	Mg	K/Mg
EUF-extract												
0– 5 min	0.57	0.32	1.8	2.17	0.18	11.8	2.94	0.11	25.8	0.82	0.04	20.0
5–10 min	1.13	0.61	1.9	3.64	0.38	9.6	3.65	0.21	17.1	1.68	0.06	26.7
10–15 min	1.10	0.57	1.9	2.85	0.33	8.6	1.96	0.19	10.4	1.13	0.06	17.6
15–20 min	0.97	0.51	1.9	2.26	0.28	8.0	1.20	0.15	8.2	0.68	0.05	12.8
20–25 min	0.87	0.48	1.8	1.91	0.26	7.3	0.74	0.11	6.7	0.49	0.05	10.9
25–30 min	0.76	0.45	1.7	1.73	0.24	7.1	0.57	0.09	6.1	0.36	0.04	9.2
30–35 min	1.06	0.71	1.5	2.14	0.40	5.4	0.55	0.14	4.0	0.41	0.06	7.1
Total	6.46	3.82	1.7	16.70	2.17	7.7	11.61	1.06	10.9	5.57	0.42	13.2
AL-extract	22.5	38.0	0.6	31.10	15.4	2.0	18.5	3.5	5.3	9.0	0.8	11.2

K to Mg released by EUF was more similar to the ratio in which plant roots remove these nutrients from the soil than was the case with AL-extraction.

As shown in Table 3, the release of both K and Mg by EUF decreased during the reaction time. However, the release of potassium decreased faster than the release of magnesium. In the very fertile clay soil (C 10) the release of both K and Mg decreased very slowly, which may prove that this soil has an ability to supply crops with both K and Mg for a long time. In the other clay soil (R 149) as well as in the sandy soils (R 154 and L 305) the release of potassium decreased much faster than the release of magnesium. This may indicate that the ability of these soils to supply plants with magnesium is more durable than their corresponding ability with respect to potassium. Even if the sandy soils (R 154 and L 305) have an immediate need of magnesium fertilization, as may be evident from the very low release of Mg, they probably will have a larger need of potassium fertilization in the future.

The results of the field experiments on the soils used in this investigation have shown that the response to potassium and magnesium fertilization varies unaccountably. Thus, there is an insignificant correlation between AL-extracted amounts of K and Mg in topsoil and the response to potassium and magnesium fertilization, respectively. However, the results of this investigation indicate that EUF-desorption may give valuable information about both the immediately and the potentially available potassium and magnesium, and hence demonstrate that the EUF-method may have an advantage over the AL-extraction in describing the potassium and magnesium status of agricultural soils.

Acknowledgements Research grants from the Swedish Council for Forestry and Agricultural Research and from the Foundation for Plant Nutrition are gratefully acknowledged. I would also like to thank Mr Bert Forsberg for assistance with the laboratory work.

References

1   Egner H, Riehm H and Domingo W R 1966 Untersuchungen über die chemische Bodenanalyse als Grundlage für Beurteilung des Nährstoffzustandes der Böden. II. Chemische Extraktionsmethoden zur Phosphor- und Kaliumbestimmung. K. Lantbrukshoegsk. Ann. 26, 199–215.
2   Frederiksson L 1961 Die Bodenuntersuchung in Schweden und ihre Auswertung. Landwirtsch. Forsch. Sonderh. 15, 74–82.
3   Hahlin M and Johansson L 1977 The ability of some methods of analysis to describe the phosphorus and potassium status in the soil. Lantbrukshoegsk. Medd. Ser. A 271, 20–33.
4   Johansson O A H and Hahlin M 1977 Potassium/magnesium balance in soil for maximum yield. Proc. Int. Seminar on Soil Environment and Fertility Management in Intensive Agriculture, Tokyo pp 487–495.
5   Németh K 1972 Bodenuntersuchung mittels Electro-Ultrafiltration (EUF) mit mehrfach variierter Spannung. Landwirtsch. Forsch. Sonderh. 27, 184–196.
6   Németh K 1976 The determination of effective and potential availability of nutrients in soil by Electro-Ultrafiltration. Appl. Sci. Dev. 8, 89–111.

A comparison of electro-ultrafiltration and quantity/intensity measurements of soil potassium with its uptake by ryegrass in Scottish soils

A. H. SINCLAIR

The Macaulay Institute for Soil Research, Aberdeen, Scotland, AB9 2QJ

Key words EUF–K Exchangeable K Fluvioglacial sand Glasshouse Lower Old Red Sandstone till Perennial ryegrass Q/I-K Quartz-mica-schist.

Summary Electro-ultrafiltration (EUF) and quantity/intensity (Q/I) parameters of soil K were compared for 14 soils from each of three soil series. The K desorbed by EUF during the first 10 min (K_{10}) was closely correlated with the equilibrium activity ratio (AR_0) for soils of the same series, but differences between series reflected the soil K-buffering capacity, indicating that K_{10} includes loosely held exchangeable K and is not strictly an intensity measurement.

EUF values were compared with conventional soil test methods for predicting K-uptake and dry-matter yield of ryegrass grown in the glasshouse. Correlation coefficients between K uptake at the first cut were 0.80 for K_{10}, 0.88 for K_a (the initially labile K derived from the Q/I curve), 0.92 for K_{35} (desorbed by EUF in 35 min) and 0.97 for K_{ex} (1.0 M ammonium acetate extraction).

Introduction

Electro-ultrafiltration (EUF) is a relatively recent development in soil testing, and it is important to evaluate it fully by comparison with conventional soil test methods. EUF using three field strengths has been suggested as a useful method for separating the K fractions which are important for plant growth[8,10,11]. Potassium quantity/intensity (Q/I) relationships are accepted for defining the soil K intensity (AR_0), labile or exchangeable K, and K-buffering power[3,4]. A linear correlation exists between the equilibrium activity ratio (AR_0) and the K extracted by EUF within a desorption time of 10 min (K_{10}) for a particular soil enriched and depleted to different K levels, but the relationship between these two parameters changes for different soils[7].

Potassium extracted by EUF has been shown to correlate more closely with K-uptake by barley plants than AR_0[7], but no account was taken of the change in AR_0 during cropping *i.e.* the soil K-buffer capacity. Previous work on Scottish soils cropped with ryegrass showed that K extracted by EUF within 35 min (K_{35}) was a better indicator of K uptake than either K_{10} or the initially labile K (K_a) obtained from the Q/I relationship, and gave a similar correlation to conventional ammonium acetate-exchangeable K (K_{ex})[15].

This paper compares EUF-extractable fractions with recognised physico-chemical parameters for soil K, with particular emphasis on the postulate that K_{10} is a measure of the intensity factor of K in soils. EUF is also compared with conventional soil test methods used to predict K availability, as measured by K-uptake and dry-matter yield of ryegrass grown in the glasshouse and with the use of soils having a wide range of clay and K contents, and with K-buffer capacity.

Materials and methods

Soils

A total of 42 cultivated soils representing three soil series were selected to give a range of parent materials, texture and clay mineralogy (Table 1). The soils were stored moist in a cool place, and passed through a 2-mm sieve, prior to chemical analyses. The ammonium acetate-exchangeable K content of the soils ranged from 1 to 20 meq kg^{-1}. K was added to some soils at rates of 0.5, 1.0 and 2.0 meq K kg^{-1} in order to follow changes in soil K measurements. Results are reported on an oven-dry basis.

EUF determinations

Sub-samples containing 5 g oven-dry soil were extracted by EUF using varied field strengths with a Vogel, Model 723 programmed instrument[15]. Amounts of K desorbed during five-minute intervals at 20°C were plotted against time to give conventional EUF graphs.

Q/I determinations

Quantity/intensity isotherms were determined for some soils from each series. Samples were gently shaken at $20 \pm 2°C$ for 0.5 h with 0.01 M CaCl$_2$ solution containing KCl from 0.0005 to 0.0035 M, using the equivalent of 5 g oven-dry soil to 50 cm^3 solution, and with 0.01 M CaCl$_2$ containing no K, at soil:solution ratios from 1:10 to 1:250[13]. The change in soil K during equilibration (ΔK) was calculated and plotted against the final activity ratio (AR^K) were

$$AR^K = \frac{a_K}{\sqrt{a_{(Ca, Mg)}}} = \frac{c_K}{\sqrt{c_{(Ca, Mg)}}} \cdot \frac{\gamma_K}{\sqrt{\gamma_{(Ca, Mg)}}}$$

and γ is the activity coefficient. The ratio $\gamma_K / \sqrt{\gamma_{(Ca, Mg)}}$ was taken as 1.18[2]. The values of AR^K at ΔK = 0 (AR$_0$) were obtained by interpolation. Equilibrium activity ratios (AR$_0$) for the remaining soils were determined from two or three points near to $\Delta K = 0$[13].

The soil K buffer capacity (BC$_0$) was measured as the tangent to the Q/I isotherm at AR$_0$.

The initially labile or exchangeable K (K$_a$) that can be removed by ryegrass was determined from the equation

$$K_a = (AR_0 - AR_e) \times BC_0$$

where AR$_e$ is the lowest activity ratio to which ryegrass can reduce the soils on intensive cropping, and is taken as 3×10^{-4} M‡ from earlier work on Scottish soils[13].

Table 1. Soils used

Series	Parent material	Symbol	Soil no.	Clay %	Texture
Boyndie	Fluvioglacial sand	BY	1–14	4–11	Sand/loamy sand
Strichen	Quartz-mica-schist	ST	15–28	17–23	Sandy loam/sandy clay loam
Laurencekirk	Lower Old Red Sandstone till	LK	29–42	19–30	Sandy loam/sandy clay loam

Exchangeable and acetic acid-extractable K

Exchangeable K (K_{ex}) in the uncropped soils was determined by 1.0 M ammonium acetate extraction[6]. Acetic acid-extractable K (K_{HAc}) was determined by shaking soil with 0.43 M acetic acid at soil:solution ratio (1 : 40 w/v).

Pot experiment

Soils were cropped in duplicate without added K in two randomised blocks of plastic pots each holding 2.8 l (3.0 − 3.5 kg) moist soil that had been passed through a 6.4-mm riddle and given, at potting, 0.33 g N and 0.36 g P as $(NH_4)_2HPO_4$, 12 mg Mg as $MgSO_4.7H_2O$ and enough $CaCO_3$ to

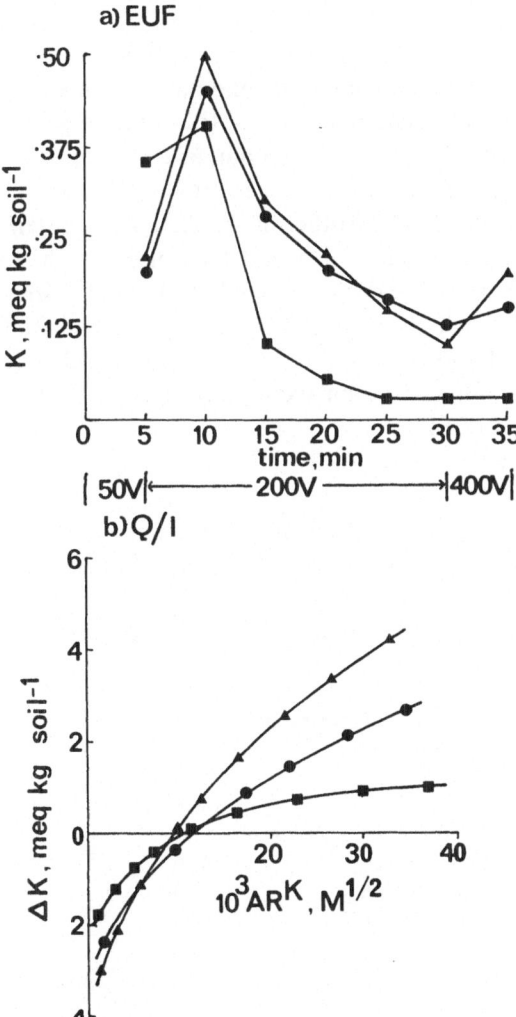

Fig. 1. Potassium EUF curves and Q/I isotherms for representative soils of the three series:
■ Boyndie; ● Strichen; ▲ Laurencekirk
a) EUF
b) Q/I

raise soil pH ($CaCl_2$) to 5.3. Each pot was sown with 0.60 g perennial ryegrass (*Lolium perenne*, cv. S23). Deionised water was added to maintain soil moisture (pF $\simeq$ 2.2). The grass was cut three times during 16 weeks. Nitrogen was added each week in a mixed solution of NH_4NO_3 and $Ca(NO_3)_2 \cdot 4H_2O$ in a 1 : 1 mole ratio of N to balance N uptake and to maintain soil pH. Uptake of Mg and S were balanced by addition of $MgSO_4 \cdot 7H_2O$.

The grass samples were dried for 48 h at 80°C and ground in a mill, before being ashed at 450°C for 16 h. K was determined by flame emission spectrometry in 0.05 M HCl extracts of the ash.

Results and discussion

EUF and Q/I plots

Typical plots of K desorbed by EUF in 5-min intervals, and quantity/intensity isotherms, are shown in Fig. 1 for representative soils of the three series. K desorption is practically complete after 25 min at a potential difference of 200 V in the soil of the sandy BY series. These sandy soils have few sites for K sorption as shown by their low buffering capacity in Q/I isotherms. The LK soils contain 19–30% clay, predominantly vermiculite with mica and kaolinite, and have higher buffering power than the BY soils. The desorption of K from the LK soils is incomplete after 30 min and further K is desorbed by raising the potential difference to 400 V (Fig. 1a). This additional K appears to come from exchange sites as the total K desorbed over 35 min (K_{35}) is on average only 54% of the 1 M ammonium acetate-exchangeable K (K_{ex}).

Relationship between K_{10} and AR_0

AR_0 is, by definition, a measurement of the chemical potential of potassium ions relative to calcium ions in the soil, independent of the concentrations and composition of the soil solution[3,4]. Potassium desorbed over 10 min (K_{10}) has been used as an empirical measure of soil K intensity[8,10,11,7] and correlates

Table 2. Linear regression equations of EUF and Q/I soil K measurements

Soil series	Linear equation	% Variance accounted for
BY	$K_{10} = 0.09 + 76\,AR_0$	93.0
ST	$K_{10} = 0.11 + 134\,AR_0$	99.0
LK	$K_{10} = 0.10 + 202\,AR_0$	99.6
BY	$K_{35} = 0.08 + 129\,AR_0$	86.9
ST	$K_{35} = 0.47 + 237\,AR_0$	97.4
LK	$K_{35} = 0.46 + 433\,AR_0$	99.0
BY	$K_{30-35} = 0.009 + 0.058\,K_{10}$	70.6
ST	$K_{30-35} = 0.040 + 0.142\,K_{10}$	92.2
LK	$K_{30-35} = 0.067 + 0.180\,K_{10}$	92.2

closely with AR_0 for soils of the same series (Table 2). However, the slope of the linear regression differs for each series and is steeper the greater the soil K-buffering capacity (Fig. 1b). This dependence on soil K-buffering capacity indicates that K_{10} cannot be a true intensity measurement. K_{35} has been considered to be a quantity parameter related to exchangeable K [8,10], but it also correlates well with AR_0 within each soil series (Table 2).

Soil K-buffer capacity

It has been suggested that K_{10} differentiates between soils of different K selectivity better than AR_0 [7], which could only be expected if K_{10} reflected buffer capacity. However, AR_0 usually decreases as buffer capacity (BC_0) increases (Fig. 1b), but within a limited range of AR_0 values BC_0 and K_{10} are in fact linearly related (Fig. 2). This confirms that K_{10} reflects soil K-buffer capacity and

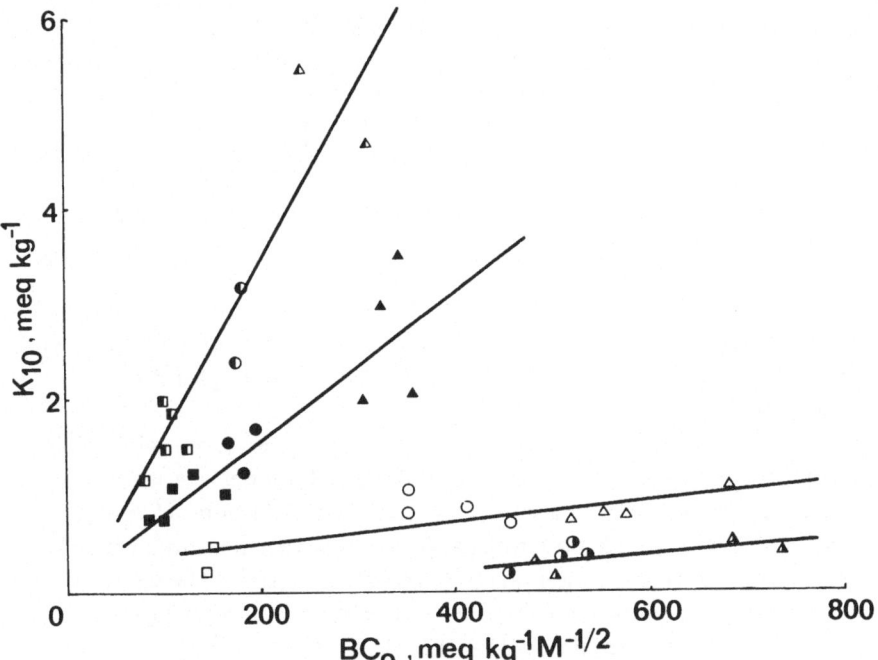

Fig. 2.   Relationship between K_{10} and soil K-buffer capacity, BC_0:
Legend

□ ■ ◨ Boyndie
◑ ○ ● ◔ Strichen
▲ △ ▲ ▲ Laurencekirk
      ◑ ▲ $AR_0 < 2 \times 10^{-3}M^{\frac{1}{2}}$
□ ○ △ $AR_0 = (2 - 8) \times 10^{-3}M^{\frac{1}{2}}$
■ ● ▲ $AR_0 = (9 - 15) \times 10^{-3}M^{\frac{1}{2}}$
◨ ◔ △ $AR_0 = (16 - 28) \times 10^{-3}M^{\frac{1}{2}}$

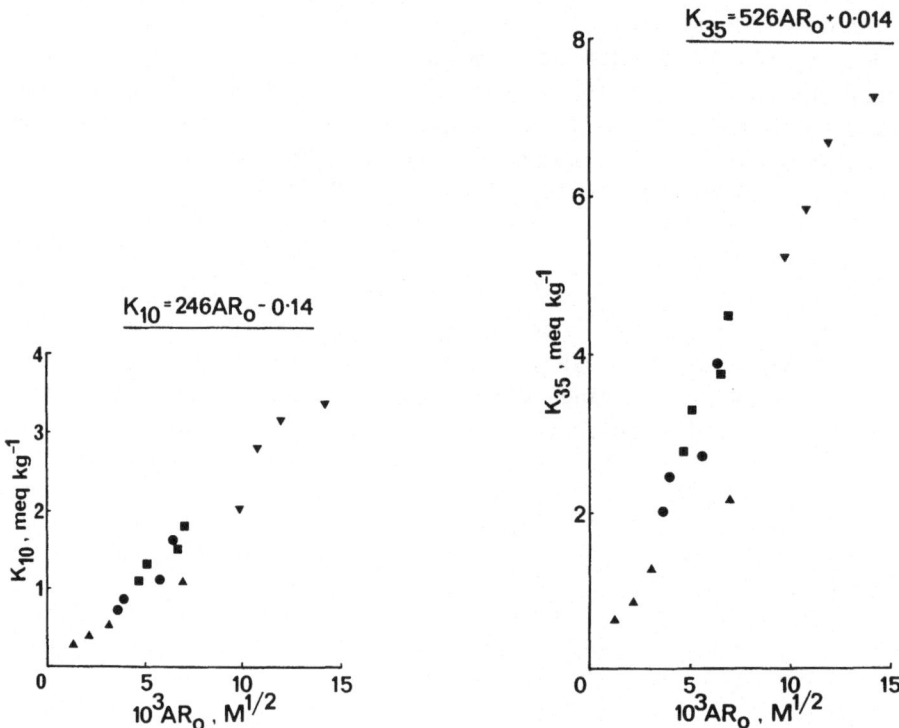

Fig. 3.   Effect of added K on EUF and AR_0 values for four soils from the Laurencekirk series.

includes loosely held exchangeable K, and is, therefore, not a true intensity parameter. For the three soil series, K_{10} averaged 43% of K_{ex} for BY, 27% for ST and 24% for LK.

The slopes of the linear regression of K_{10} and K_{35} on AR_0 (Table 2) differentiate between soil series of different K-buffer capacity. Similar relationships for soils within a series also holds with addition of 0.5, 1.0 and 2.0 meq K kg^{-1} (Fig. 3). Soil K buffering has previously been measured with EUF by the K desorbed during the interval 30–35 min (K_{30-35})[8, 10]. The values increase with increasing K_{10} within a soil series (Table 2), and for a given value of K_{10}, K_{30-35} increases in order BY < ST < LK (Fig. 1a).

Both K_{30-35} and K_{10} decreased rapidly with removal of K by cropping in the glasshouse. At the first cut, mean values of K_{10} and K_{30-35} were 0.38 and 0.11 meq kg^{-1}, respectively, compared to 1.61 and 0.25 meq kg^{-1} before cropping. Unlike K_{30-35}, linear buffer capacity (LBC), given by the slope of the near-linear upper part of Q/I isotherms, has been shown to be virtually unaltered by intensive cropping or addition of K[13, 1]. Thus, for the soils studied here, K_{30-35} does not appear to indicate the rate of K supply from the interlayers of clay minerals, as reported in earlier work[9].

K_{30-35} is a desorption parameter whereas LBC is often measured over the

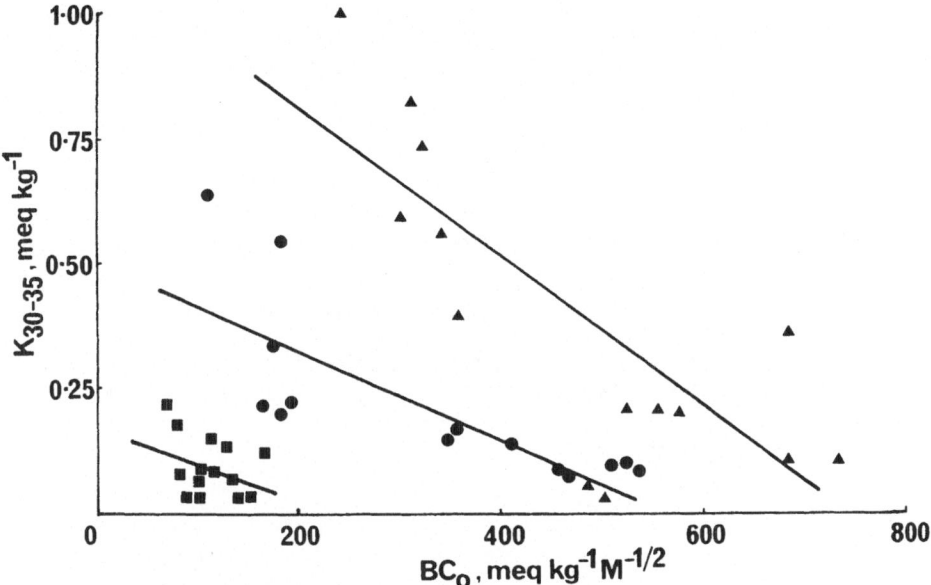

Fig. 4. Relationship between buffer capacity parameters K_{30-35} and BC_0: ■ Boyndie; ● Strichen; ▲ Laurencekirk

adsorption part of the Q/I isotherm. The buffer capacity term BC_0 may be more closely related to K_{30-35} than is LBC, and as BC_0 increases K_{30-35} tends to decrease within a soil series (Fig. 4). This trend can be explained by the dependence of K_{30-35} on K_{10} (Table 2), and on the increase in BC_0 with decreasing AR_0 when values lie on the curved part of the Q/I. Thus, K_{30-35} does not appear to measure the change in K intensity on removal or addition of K in the same way as BC_0.

Correlations between K_{30-35} and non-exchangeable K have been obtained[12], but in other work, K_{30-35} has been related to the release of initially non-exchangeable K in close equilibrium with the exchangeable pool, but not the release of K from the interlayers of micaceous clay minerals[15]. The mobility of this interlayer K, where diffusion coefficients ranged from 10^{-22} to 10^{-20} cm^2s^{-1} (ref.[14]), is considered to be too low to contribute significantly to K desorption at 20^{0}[5]. Thus, EUF at 20^0, as well as Q/I isotherms, appears only to extract fractions of immediately exchangeable K.

Soil K and uptake of K by ryegrass

K-uptake by ryegrass at the first cut (six weeks) is compared with soil test methods in Fig. 5. The coefficient of variation of cumulative K-uptake was 5% for the first cut, 3% for the second and 2% for the third. Six soils, where K uptake appeared to have reached a maximum that was not limited by soil K, are omitted.

Table 3. Correlation coefficients of linear regressions of measurements of soil K availability on cumulative K uptake and yield of ryegrass at different cropping stages

K measurement		K uptake at cut no			Yield at cut no		
		1	1 + 2	1 + 2 + 3	1	1 + 2	1 + 2 + 3
EUF	K_{10}	0.80***	0.77***	0.74***	0.39*	0.58***	0.61***
	K_{35}	0.92***	0.90***	0.88***	0.50**	0.72***	0.77***
Q/I	AR_0	0.49**	0.43**	0.39*	0.21	0.31	0.29
	K_a	0.88***	0.87***	0.85***	0.47**	0.72***	0.77***
lab	K_{ex}	0.97***	0.98***	0.98***	0.51**	0.76***	0.86***
	K_{HAc}	0.81***	0.82***	0.79***	0.50**	0.71***	0.72***

*$P < 0.05$; **$P < 0.01$; ***$P < 0.001$

Linear correlation coefficients (Table 3) show that K_{ex} correlates more closely with K-uptake at each cut than values obtained by the other methods.

In previous work with ten Scottish soils[15], both K_{35} and K_{ex} correlated equally well with K-uptake by ryegrass. In the present study, using soils with a wide range of clay and K contents and K-buffer capacity, both EUF and Q/I measurements overestimate K-uptake especially for the BY series (Fig. 5), an overestimation that increases with continuous cropping (Table 3). The poorest correlation with K-uptake is obtained with AR_0 values because these measurements do not take into account the rate of decrease of AR_0 during cropping. AR_0 values of the BY series, with lowest buffer capacity, can be expected to decrease more than those for the ST and LK series, and hence AR_0 measured before cropping overestimates K-uptake. When the soil K-buffering term, BC_0, is taken with AR_0 and used to calculate K_a, the amount of exchangeable K removed by ryegrass, the correlation with K-uptake is greatly improved (Fig. 5d). Owing to the curvature of the lower part of the Q/I isotherms, K_a may be underestimated in some soils because BC_0 may increase during cropping, depending on the position of AR_0 on the Q/I isotherm before and after cropping. The predictive value of K_a is further reduced when initially non-exchangeable K becomes available because Q/I isotherms do not include non-exchangeable K, and the rate of release is not necessarily proportional to the amount of exchangeable K[14].

K_{10} accounts for 64% of the variation in K-uptake at the first cut, which is more than AR_0, probably because K_{10} also reflects buffer capacity (Fig. 2), but again K_{10} values of BY soils overestimate K-uptake. When more K is extracted by EUF, i.e. K_{35}, the correlation with K-uptake improves.

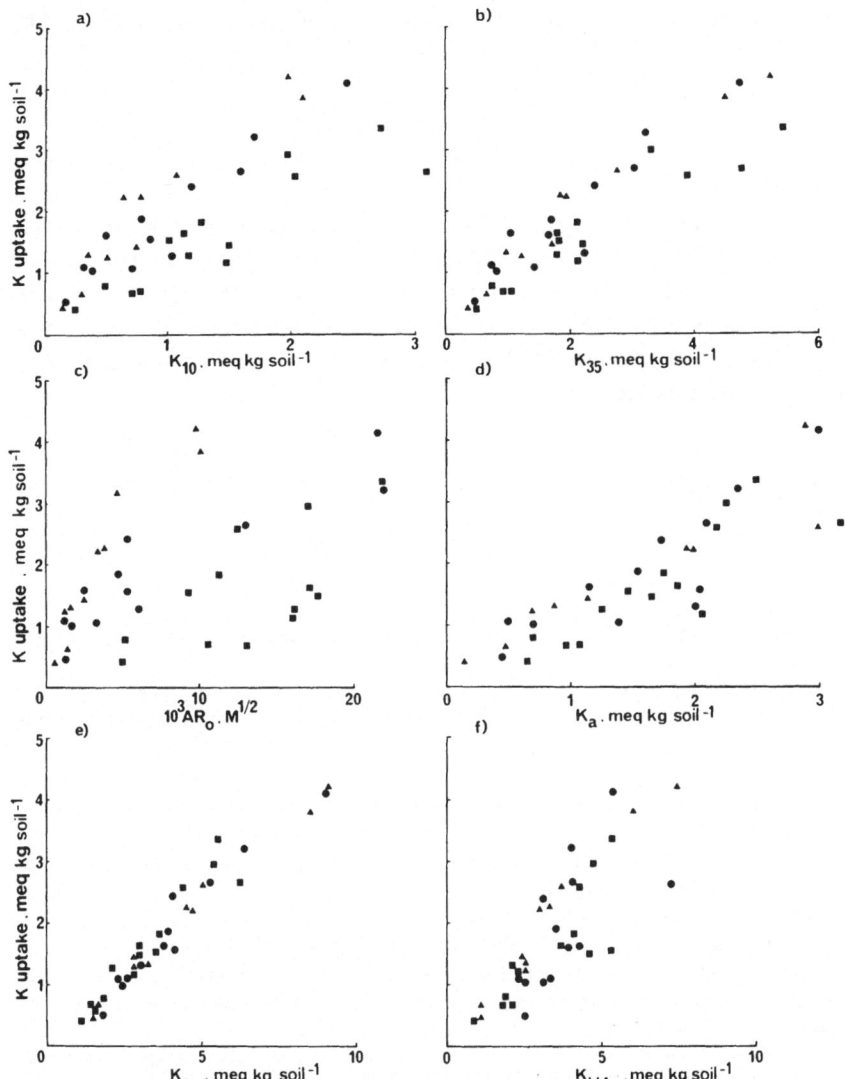

Fig. 5.   Relationship between K uptake at the first cut of ryegrass and soil K measurements: a) K_{10}; b) K_{35}; c) AR_0; d) K_a; e) K_{ex}; f) K_{HAc}:
■ Boyndie; ● Strichen; ▲ Laurencekirk

Soil K and yield of ryegrass

The coefficient of variation of cumulative yield was 6% for the first cut, and 5% for the second and third cuts. K_{ex}, K_{35}, K_a and K_{HAc} accounted for a similar amount of the variation in yield and more than AR_0 and K_{10} (Table 3). AR_0 accounted for least variation, as expected, because buffer capacity is not taken into account, while K_{10} was better than AR_0 because K_{10} is partly regulated by soil K-buffer capacity.

Conclusions

EUF measurements K_{10} and K_{35} correlated well with K-uptake by ryegrass. However, because K_{10} is related to K-buffer capacity it should not be regarded purely as an intensity measurement. Quantity/intensity parameters also correlated well with K-uptake provided the K-buffer capacity is taken into account. But neither Q/I nor EUF measurements correlated with K-uptake by ryegrass as closely as did ammonium acetate-exchangeable K.

Acknowledgements I wish to thank Mrs R G McPherson, Mrs G Coutts and members of the Departments of Soil Fertility, Spectrochemistry and Statistics of this Institute for technical assistance.

References

1   Addiscott T M 1970 The potassium Q/I relationships of soils given different manuring. J. Agric. Sci. Camb. 74, 131–137.
2   Beckett P H T 1965 Activity coefficients for studies on soil potassium. Agrochimica 9, 150–154.
3   Beckett P H T 1967 Potassium potentials *In* Soil Potassium and Magnesium. Techn. Bull. Ministry of Agriculture, Fisheries and Food 14, 32–37.
4   Beckett P H T 1972 Critical cation activity ratios. Adv. Agron. 24, 379–412.
5   Grimme H 1980 The effect of field strength on the quantity of K desorbed from soils by electro-ultrafiltration. Z. Pflanzenernaehr. Bodenkd. 143, 98–106.
6   Jackson M L 1958 Exchangeable metallic cation determinations for soils. *In* Soil Chemical Analysis. Constable, London pp 82–110.
7   Nair P K R and Grimme H 1979 Q/I relations and electroultrafiltration of soils as measures of potassium availability to plants. Z. Pflanzenernaehr. Bodenkd., 142, 87–94.
8   Németh K 1972 The determination of desorption and solubility rates of nutrients in the soil by means of electroultrafiltration (EUF). Proceedings of the 9th Coloquium of the International Potash Institute. pp 171–180.
9   Németh K 1975 The effect of K fertilization and K removal by ryegrass in pot experiments on the K concentration of the soil solution of various soils. Plant and Soil 42, 97–107.
10 Németh K 1976 The determination of effective and potential availability of nutrients in the soil by means of electroultrafiltration. Appl. Sci Developm. 8, 89–111.
11 Németh K 1979 The availability of nutrients in the soil as determined by electro-ultrafiltration (EUF). Adv. Agron. 31, 155–188.
12 Németh K and Forster H 1976 Beziehungen zwischen Ertrag und K-entzug von Ackerbohnen (*Vicia faba*) sowie verschiedenen K-fraktionen von Böden. Bodenkultur 27, 117–119.
13 Sinclair A H 1979 Availability of potassium to ryegrass from Scottish soils 1. Effects of intensive cropping on potassium parameters. J. Soil Sci. 30, 757–773.
14 Sinclair A H 1979 Availability of potassium to ryegrass from Scottish soils II. Uptake of initially non-exchangeable potassium. J. Soil Sci. 30, 775–783.
15 Sinclair A H 1980 Desorption of cations from Scottish soils by electro-ultrafiltration. J. Sci Food Agric. 31, 532–540.

EUF-K as measure of K availability index for Tamil Nadu soils (India)

K. M. RAMANATHAN
Department of Soil Science and Agricultural Chemistry Tamil Nadu Agricultural University, Coimbatore, India

and K. NÉMETH
Büntehof Agricultural Research Station, Hannover, Federal Republic of Germany

Key words K potential K reserve EUF-K K uptake by rice

Summary Representative soils of Tamil Nadu could be grouped into three categories based on the cumulative EUF desorption K curves which were exponential and displayed marked differences in the magnitude of K release by the soils employed. The cumulative K desorption values and the cumulative K uptake values of rice were found to be closely correlated. When compared with any other method, the cumulative EUF desorption K was found to be a better measure of K availability index.

Introduction

Although much has been said and written to unravel the various factors affecting K availability, yet not a single method has been found to be universally applicable to assess the K available status of soils. The heterogeneity of a soil coupled with the dynamic equilibrium of K among the forms in which it exists in the soil makes the problem rather complex.

Numerous methods have been advocated by several workers to measure the available K status of soils under different situations[1,2,3,4,6,8,10,12,14,16,17,18]. Methods of K availability can be meaningfully employed only when a close relationship exists between K extracted and K absorbed by the plant. Such a method should indicate the immediately available K in the soil (intensity factor), the reserve K (quantity- or capacity factor), and the degree of passage of K through the soil solution (rate factor). EUF is considered to be one of the methods capable of determining the concentration of K in the soil solution, the exchangeable K and the degree of capacity of a soil to maintain the K potential by the release of reserve K into the soil solution[7,13,14].

In this paper comparisons are made of the relative efficiencies of different methods of K availability indices for Tamil Nadu soils as reflected by the dry matter yield and K uptake of rice.

Materials and methods

Laboratory experiments

Twenty-five representative surface (0–15 cm) soil samples collected from all districts of the State of Tamil Nadu (India) were used in this study. K was extracted using different extractants viz. water, N NH_4 OAc, 0.1 N HCl, 0.5 N HCl, 0.1 N HNO_3, 0.5 N HNO_3, 0.5 N EDTA, 0.5 N NaCl, 1 per cent citric acid, 0.01 M $CaCl_2$ and N HNO_3. The K potential parameters AR_0, $-\Delta K^0$, PBC^K, free energy changes ΔG, and K releasing power of soils by stepwise extraction with 0.01 N HCl[2,6,20] were also determined. The EUF–K was determined using the EUF apparatus of the Büntehof Agricultural Research Station, Hannover, Germany. The EUF apparatus and the principles of the technique involved have been discussed in detail by Németh[12].

The amounts of K extracted in 5-minute intervals at 20°C and 200 V were summed up and the cumulative EUF–K values were plotted against desorption time in minutes to obtain cumulative desorption K curves.

Table 1. Physico-chemical characteristics of soils

Soil No.	Location	Clay %	Silt %	CEC meq/ 100 g	pH	Organic carbon %	CaCO$_3$ %
1	Kanchipuram	15.1	6.6	10.2	8.0	0.39	1.00
2	Sriperumpudur	19.1	6.5	6.8	6.9	0.15	0.20
3	Coimbatore	38.2	8.2	28.8	8.2	0.60	2.35
4	Vadavalli	18.0	5.8	13.8	8.0	0.30	1.80
5	Kaveripattanam	3.4	6.7	6.1	7.8	0.06	0.20
6	Dharmapuri	38.2	14.3	35.2	7.2	0.19	1.15
7	Kalkulam	14.6	3.4	5.3	4.7	0.54	–
8	Agasteeswaram	14.9	1.6	4.9	5.5	0.27	–
9	Tirumangalam	39.0	7.5	40.9	7.9	0.30	1.75
10	Melur	37.2	7.9	4.7	7.5	0.15	0.30
11	Gudiyatham	11.3	7.6	29.5	8.0	0.45	2.20
12	Tiruvannamalai	8.2	7.0	5.0	7.2	0.03	0.25
13	Paramakudi	21.9	12.9	20.2	6.7	0.39	0.80
14	Mudukulathur	20.9	5.1	14.2	7.6	0.33	0.80
15	Tiruchengode	13.8	2.8	16.0	6.5	0.33	0.70
16	Attur	11.8	22.7	21.9	7.8	0.39	2.35
17	Cuddalore	13.3	9.5	10.9	7.0	0.36	0.45
18	Tindivanam	23.1	10.7	10.3	6.7	0.24	0.30
19	Aduthurai	52.4	9.3	27.3	6.7	0.48	1.00
20	Gantharvakottai	14.8	7.5	2.8	5.8	0.24	–
21	Ambasamudram	15.8	3.2	8.0	5.4	0.27	0.30
22	Kovilpatti	43.0	13.1	58.5	8.0	0.18	1.90
23	Musiri	17.8	14.0	17.1	7.9	0.30	2.10
24	Tiruchy	17.8	29.4	26.9	7.4	0.51	1.10
25	Nanjanad	15.8	5.5	19.8	5.6	5.28	–

Experiments with plants

A pot experiment was conducted to study the K-supplying power of these 25 soils by relay cropping with rice, variety IR 25, as test crop. Six hundred-g samples of soil that had passed through a 2-mm sieve were transferred to plastic pots lined with a layer of wax. To promote good growth quantities of N and P (50 ppm N and 25 ppm P) were added to every crop. Fifty well sprouted IR 26 rice seedlings were planted each time, and allowed to grow for 30 days. Deionised water was added every day to maintain a uniform level of submergence.

The plants were pulled out carefully after 30 days of growth, the roots were washed and the adhering soil was put back into the pots. The plants were dried and the dry-matter yields recorded. A total of 6 crops of IR 26 rice were thus raised. After each harvest the plant samples were analysed for their K contents, with the use of a standard procedure[9]. The cumulative dry-matter yields and the cumulative K uptake values were calculated. Simple correlations were worked out between various K availability indices, on the one hand, and the values on cumulative dry-matter yield and cumulative K uptake, on the other hand.

Results and discussion

On account of their variability in origin from different geological formations, like Archaean, Tertiary and Quarternary, and on account of their occurrence in

Table 2. Ranges of potassium extracted by different extractants from the 25 soils used in the experiment

Method No.	Name of the extractant/method	Ranges of values for 25 soils	Mean values
1	H_2O ppm	4 – 48	16
2	N NH_4OAc ppm	23 – 460	153
3	0.1 N HCl ppm	35 – 330	110
4	0.5 N HCl ppm	40 – 360	126
5	0.1 N HNO_3 ppm	33 – 300	107
6	0.5 N HNO_3 ppm	35 – 455	153
7	0.5 N EDTA ppm	33 – 250	90
8	0.5 N NaCl ppm	25 – 193	70
9	1 per cent citric acid ppm	35 – 300	102
10	0.01 M $CaCl_2$ ppm	35 – 295	96
11	N HNO_3 ppm	100 – 1950	667
12	Non-exchangeable K ppm	72 – 1490	494
13	0.01 N HCl cumulative K release ppm	85 – 680	244
14	Total K (per cent)	0.79 – 2.16	1.38
15	EUF–K cumulative desorption (mg/100 g)	1.20 – 27.9	8.56
	K-potential parameters		
16	$AR_0^K \times 10^{-3}((m/1)^{\frac{1}{2}})$	2.1 – 20.4	6.9
17	$-\Delta K^0(meq/100\,g)$	0.11 – 1.36	0.47
18	PBC^K	15.9 – 267.7	76.3
19	$\Delta G = 1364 \log \dfrac{a_K}{\sqrt{a_{Ca}}}$	3688 – 2328	3088

different agroclimatic regions of the State, the physico-chemical characteristics
of the soils revealed a wide heterogeneity (Table 1).

Potassium status of soils

The ranges of values of K extracted by different extractants for the different
soils are presented in Table 2. These data also indicate clearly the heterogeneity of
the soil with wide variations in K status.

Cumulative K desorption – and cumulative K uptake curves

The cumulative K desorption curves constructed by plotting the cumulative
EUF–K values against the successive desorption time in minutes are presented in

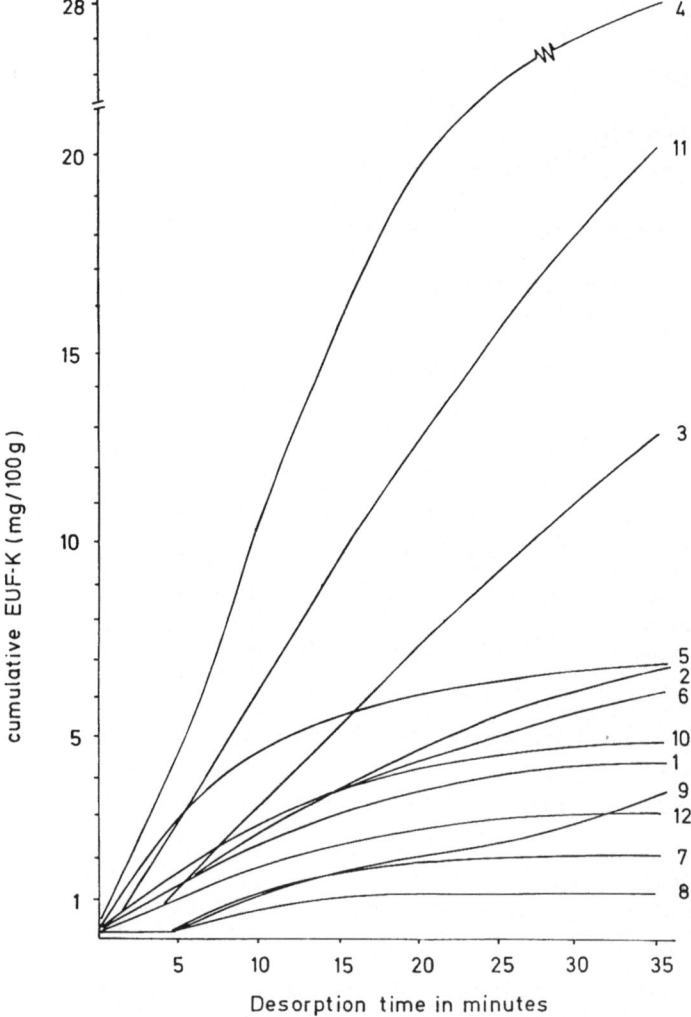

Fig. 1.   Cumulative K desorption curves of the analysed soils (Nos. 1–12).

Figs. 1 and 2. The cumulative K uptake curves of rice grown on these soils are presented in Figs. 3 and 4. These curves conformed to Cobb-Douglas exponential-type equations and were distinctly different for different soils. The best fitting exponential equations ($Y = ax^b$) of these curves are presented in Table 3.

K desorption curves

The magnitudes of the initial steepness of the curves varied markedly for different soils. Similarly the transitional as well as the flattening regions of the curves depicted interesting differences in K desorption pattern. Based on the size

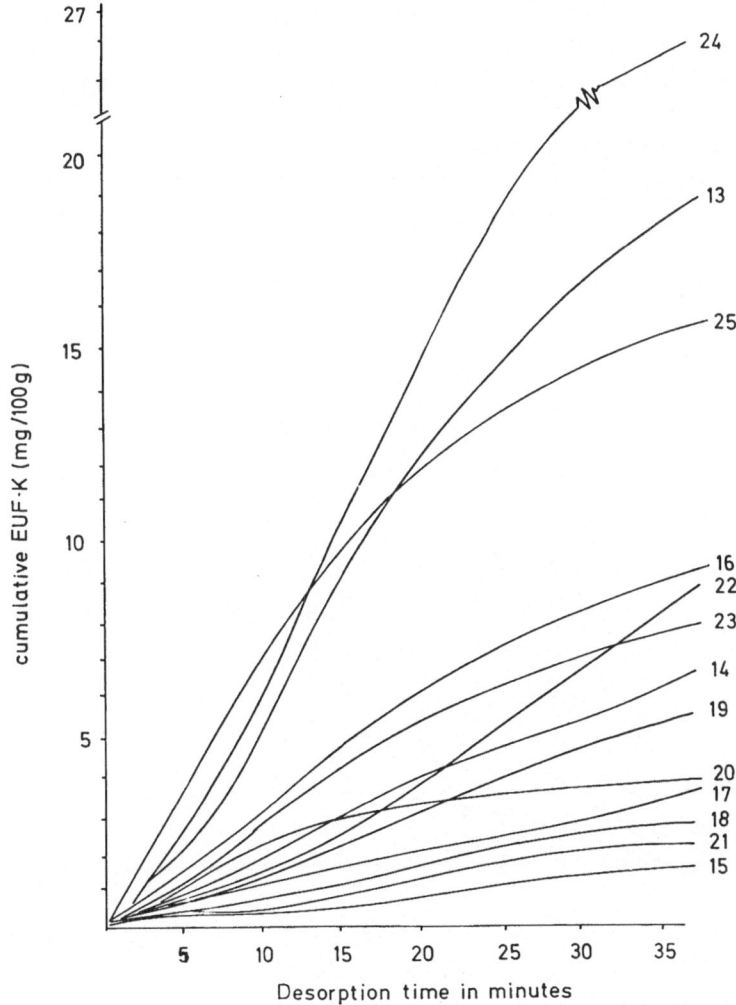

Fig. 2. Cumulative K desorption curves of the analysed soils (Nos. 13–25).

of the slope of the curves, the soils could be grouped into three categories (Figs. 1 and 2) *viz* category one consisting of soils 3, 4, 11, 13, 24 and 25, category two comprising soils 1, 2, 5, 6, 10, 14, 16, 19, 22 and 23 and category three covering the rest of the soils (7, 8, 9, 12, 15, 17, 18, 20 and 21).

The first category with a high degree of steepness in the early stages of the desorption time represented soils with a rapid rate of K desorption although the total desorbable K at the end of 35 minutes varied significantly. The high maximally desorbable K amounts clearly indicate that these soils can release sizeable quantities of K. The rapid K desorption rates for these soils reveal that K is loosely held and hence can be released with relatively little energy[15].

Such soils can be considered to have favourable intensity and replenishment characteristics (except soil No. 25). This was also revealed by the high 'a' or 'b' coefficient values of the Cobb-Douglas exponential equations obtained for these curves (Table 3). For the Nilgiris soil (No. 25-laterite), the K desorption rate rapidly fell off which was also clearly reflected in the pertinent cumulative K uptake curve (Fig. 3).

The second category with moderate slopes of the desorption curves represented soils having moderate quantities of desorbable K. This characteristic could be attributed to the combined effects of fairly low intensity factors and

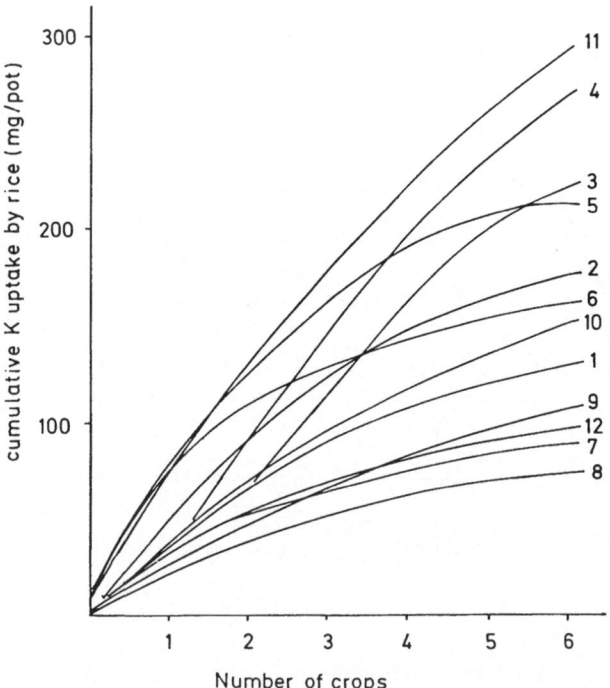

Fig. 3.   Cumulative K uptake by rice by successive cropping (Soil Nos. 1–12).

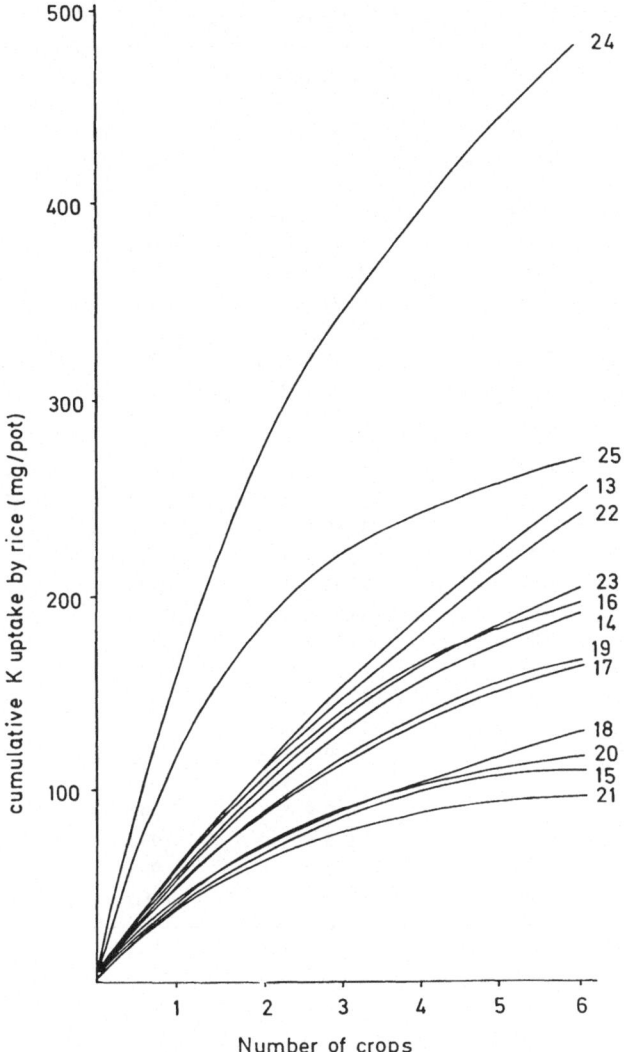

Fig. 4. Cumulative K uptake by rice by successive cropping (Soil Nos. 13–25).

strong adsorption requiring a relatively large expenditure of energy for K desorption.

The third category of curves with gentle slopes represents soils having low quantities of desorbable K. It could be hypothesised that in such soils both intensity and replenishment characteristics are quite weak. The cumulative K uptake values recorded for these soils (Figs. 3 and 4) are reflecting these characteristics.

The cumulative K uptake curves (Fig. 3 and 4) interestingly followed almost the same pattern in their steepness and magnitude as that of the cumulative K

Table 3. Cobb-Douglas exponential equations ($Y = ax^b$) for the curves obtained

Soil No.	Cumulative EUF–K curve		Cumulative K uptake curve for rice	
	ax^b values	Correlation coefficient	ax^b values	Correlation coefficient
1	0.437 × 0.692	0.985	40.42 × 0.65	0.997
2	0.241 × 0.984	0.984	58.47 × 0.62	0.987
3	0.168 × 1.250	0.996	32.47 × 1.12	0.996
4	1.321 × 0.890	0.988	41.77 × 1.09	0.992
5	1.825 × 0.395	0.978	79.77 × 0.58	0.994
6	0.254 × 0.933	0.988	79.38 × 0.40	0.998
7	0.281 × 0.582	0.982	35.57 × 0.52	0.993
8	0.104 × 0.705	0.993	24.59 × 0.64	0.991
9	0.064 × 1.263	0.724	25.35 × 0.80	0.995
10	0.661 × 0.600	0.978	38.56 × 0.78	0.997
11	0.499 × 1.069	0.993	75.71 × 0.76	0.998
12	0.353 × 0.648	0.984	34.88 × 0.59	0.997
13	0.321 × 1.189	0.987	58.75 × 0.82	0.999
14	0.130 × 1.124	0.992	47.32 × 0.80	0.996
15	0.010 × 1.454	0.996	39.76 × 0.57	0.995
16	0.270 × 1.025	0.991	57.67 × 0.71	0.993
17	0.104 × 0.998	0.998	44.61 × 0.74	0.993
18	0.024 × 1.382	0.997	35.98 × 0.72	0.994
19	0.070 × 1.263	0.998	45.77 × 0.72	0.997
20	0.407 × 0.665	0.979	35.79 × 0.67	0.993
21	0.025 × 1.290	0.996	29.82 × 0.66	0.992
22	0.020 × 1.736	0.998	51.80 × 0.88	0.998
23	0.256 × 0.996	0.987	50.55 × 0.79	0.995
24	0.350 × 1.243	0.994	173.20 × 0.58	0.996
25	1.260 × 0.735	0.984	121.12 × 0.47	0.982

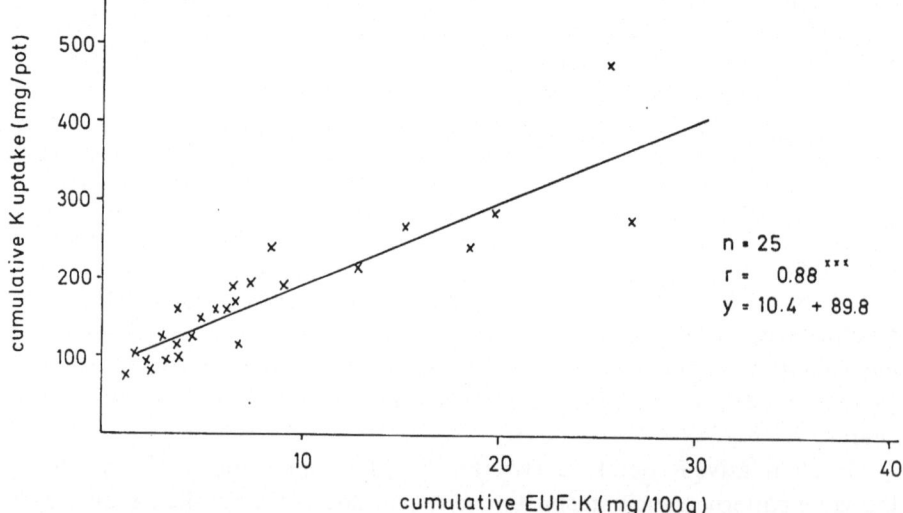

Fig. 5. Relationship between cumulative EUF–K and cumulative K uptake.

Table 4. Numerical values and levels of statistical significance of coefficients of correlation between different K availability indices (X) and cumulative dry matter yield of rice (Y_1)/cumulative K uptake (Y_2) by rice

K-availability index (X)	Correlation coefficient (r)	
	Cumulative dry matter yield (Y_1)	Cumulative K uptake (Y_2)
Water-soluble K	0.514**	0.631***
N NH$_4$OAc K	0.397*	0.807***
0.1 N HCl K	0.587**	0.824***
0.5 N HCl K	0.522**	0.840***
0.1 N HNO$_3$ K	0.666***	0.854***
0.5 N HNO$_3$ K	0.425**	0.815***
0.5 N EDTA K	0.539**	0.779***
0.5 N NaCl K	0.568**	0.793***
1 per cent citric acid K	0.535**	0.810***
0.01 M CaCl$_2$K	0.509**	0.750***
N HNO$_3$ K	0.222	0.747***
Non-exchangeable K	0.150	0.682***
Cumulative K release	0.403*	0.803***
AR$_0^K$	0.395	0.498*
$-\Delta K^o$	0.580	0.856***
PBC^K	0.152	0.389
ΔG	0.378	0.464*
Total K	0.067	0.511**
Cumulative EUF–K	0.519**	0.880***

* significant at p 0.05; ** significant at p 0.01; *** significant at p 0.001

desorption curves (Figs. 1 and 2) of the respective soils. This clearly reveals the existence of a very close relationship ($r = 0.880$***) between values of cumulative K uptake and cumulative K desorption (Fig. 5 and Table 4).

References

1 Barrow N J 1966 Nutrient potential and capacity. Aust. J. Agric. Res. 17, 849–861.
2 Beckett P H T 1964 Studies on soil potassium II. The immediate Q/I relationships of labile K in the soil. J. Soil Sci. 15, 9–23.
3 Ekpete D M 1972 Comparison of methods of available K assessment for eastern Nigeria soils. Soil Sci. 113, 213–221.
4 Elsokkary I H 1973 Evaluation of K availability indices and K release of some soils of Egypt. Potash Rev. Sub. 5, Suite 35.
5 Garman W L 1957 Potassium release characteristics of several soils from Ohio and New York. Soil Sci. Soc. Am. Proc. 21, 52–58.

6   Graham E R and Kampbell D H 1968 Soil K availability and reserve as related to the isotopic
     pool and Ca exchange equilibria. Soil Sci. 106, 101–106.
7   Grimme H 1976 Soil factors of K availability. Potassium in soils, crops and fertilizers. Bull. 10.
     Indian Soc. Soil Sci. New Delhi 141–163.
8   Hunter A H and Pratt P F 1957 Extraction of K from soils by sulphuric acid. Soil Sci. Soc. Am.
     Proc. 21, 595–598.
9   Jackson M L 1973 Soil Chemical Analysis. Prentice Hall of India, New Delhi.
10   MacLean A J 1960 Water soluble K, per cent K-saturation and pK $- \frac{1}{2}$p (Ca + Mg) as indices
     of management effects on K status of soils. Trans. 7th Int. Congr. Soil Sci. Madison 111, 86–91.
11   Nair P K R and Grimme H 1979 Q/I relations and EUF of soils as measures of K availability to
     plants. Z. Pflanzenernaehr. Bodenkd. 142, 87–94.
12   Németh K 1972 The determination of desorption and solubility rates of nutrients in the soil by
     means of EUF. Potassium in soil. Proc. 9th Colloq. Int. Potash Inst. Landshut 171–180.
13   Németh K 1978 Limitations of present soil test interpretation for K and suggestions for
     modification. A European experience. Potassium in soils and crops. Proc. Symposium Potash
     Res. Inst. India New Delhi 95–111.
14   Pratt P F 1951 Potassium removal from Iowa soils by greenhouse and laboratory procedures.
     Soil Sci. 72, 107–117.
15   Ramanathan K M 1975 Potassium release characteristics of certain soils of Tamil Nadu.
     Madras Agric. J. 62, 1–9.
16   Ramanathan K M 1977 Studies on dynamics of soil potassium. Ph. D. thesis. Tamil Nadu
     Agric. Univ. India.
17   Ramanathan K M 1978 An evaluation of K availability indices of some soils of South India. J.
     Indian Soc. Soil Sci. 26, 198–202.
18   Standford S and English L 1949 Use of flame photometer in rapid soil tests for K and Ca.
     Agron. J. 41, 446–447.
19   Thiagalingam K and Grimme H 1976 The evaluation of the K status of some Malaysian soils
     by means of EUF. Planter. Kuala Lumpur. 52, 83–89.
20   Woodruff C M and McIntosh J L 1960 Testing soils for K. Trans. 7th Int. Congr. Soil Sci. 3,
     80–84.

Application of the EUF procedure in grape production

J. EIFERT*, M. VÁRNAI and L. SZÖKE

Research Institute for Viticulture and Enology, Budapest, Hungary

Key words EUF–K EUF–Mg EUF–Mn EUF–P Grape varieties Nutrient availability Plant analysis Soil analysis

Summary The relationships between the nutrient contents in vine leaves and grape yield on the one hand and the nutrient contents in the soil on the other are described on the basis of results from pot and field experiments conducted over several years. The soils were analysed by means of conventional methods as well as by electro-ultrafiltration (EUF). The following results were obtained:

The application of high K and P fertilizer amounts in pot experiments increased the availability of Mg, Mn and Fe. A marked rise in the Mg, Mn and Fe contents was observed in the vine leaves. The exchange processes due to fertilizer addition were well indicated by the EUF method, whereas the results obtained by extraction with ammonium lactate (AL) and $CaCl_2$ were unsatisfactory.

Close and highly significant correlations were found between the EUF–P, EUF–K and EUF–Mg contents on the one hand and the P, K and Mg contents in vine leaves on the other. A close correlation also exists ($r = 0.91$***) between grape yield and EUF–K contents.

Grape yield increases with increasing EUF–K values up to 25 mg/100 g/30 min at 20°C in pot experiments (30 cm rooting depth) and only up to 12 mg/100 g/30 min at 20°C in field experiments. The soil in this field experiment had, however, EUF–K values of 12 mg/100 g in the topsoil as well as in the subsoil. When assessing the limit values, it is therefore important to consider the depth of the horizon in which the nutrients are present.

After addition of very high amounts of P fertilizer the P contents in vine leaves markedly decrease after one year, as there is a decline in the availability of phosphates. The EUF–P values measured immediately after the application of high doses of P fertilizer can only characterize the P supply status of a soil for a period of one or two years.

The P availability after application of different forms of phosphate fertilizers (superphosphate, hyperphosphate) is well indicated by means of EUF, but not by means of the AL method.

When assessing the required K and Mg values in the soil (whatever the method) the vine variety has also to be taken into account, whereas the utilization of soil phosphates depends less on the varietal differences.

Introduction

Provided that the main conditions of permanent and high yield production are given (variety, shape of vine-stock, loading, pest and disease control *etc.*), the effective fertilization will be the determining factor for production.

The main features of intensive fertilization are[3]: '*pool*' fertilization, applied before planting and '*maintenance*' fertilization, applied annually in bearing vineyards.

The amount to be applied in 'pool' fertilization can primarily be diagnosed by soil analysis (P, K, Ca, Mg), whereas plant analysis plays an important role in

* Present address: Dr. J. Eifert, Bérc u. 7, H-1016 Budapest, Hungary.

'maintenance' fertilization. The amount of 'maintenance' fertilizer depends on the yield quantity and the nutrient dynamics in the soil. Whether the annual fertilizer amounts are sufficient for the respective yield production or not, can be determined by plant analysis. In this connection soil analysis should only be used, when the method is suitable to give precise information on the nutrient dynamics in the soil.

Since 1965 simultaneous soil and plant analysis has become general practice in Hungary. In many cases, however, no correlations were found between the data of soil analysis and the results of plant analysis [2]. Intensive studies in vineyards revealed that the deficiency symptoms of certain nutrients are better identified by plant analysis than by chemical soil analysis. Particularly in soils rich in clay a good K and Mg supply was often diagnosed by soil analysis, whereas the vine suffered from K and Mg deficiency. It was, therefore, attempted to test new methods of soil analysis in pot and field experiments.

It was, moreover, intended to work out the minimum values required for optimal plant nutrition in plant as well as soil analysis, in the latter not only for the topsoil (0–30 cm), but also for the subsoil.

Material and methods

Pot and field experiments were carried out with three different soils. Some of the soil properties are given in Table 1. Besides conventional methods of soil analysis (AL extraction for K_2O and P_2O_5 and extraction with 0.025 M $CaCl_2$ for Mg) the soil samples were investigated by means of EUF as follows: 0–30 min 200 V at 20°C and 30–35 min 400 V at 80°C [4]. The samples for plant analysis were collected at flowering and ripening and analysed for the nutrients Ca, K, Mg, Mn and P.

Experiments
1. Pot experiment No. 1 was carried out with soil No. 1 which received different amounts of K (0–3800 kg/ha) and P (0–2080 kg/ha) fertilizer. Determinations were made throughout the vegetation period by soil-as well as by plant analysis.
2. A field experiment was conducted with soil No. 2 which was treated with different amounts of K_2O (0–5000 kg/ha) and P_2O_5 (0–1700 kg/ha) prior to planting. Part of the fertilizer was applied at a depth of 31–60 cm and part at a depth of 0–60 cm.
3. A pot experiment with soil No. 3 was carried out to study the effect of different forms of phosphate fertilizer.

Table 1. Some properties of the experimental soils

Soil no.	Depth (cm)	pH (H_2O)	Org. m. (%)	Clay (%)	AL–P_2O_5	AL–K_2O (mg/100 g)	$CaCl_2$-Mg
1	0–30	6.3	1.9	45	8.5	28.0	–
2	0–30	6.7	2.4	45	14.5	36.8	35.4
	31–60	6.6	1.7	45	2.6	28.8	36.0
3	0–30	5.2	1.2	50	0.4	35.0	–

Fertilizer treatments: 3150 kg K_2O/ha/30 cm
                                3847 kg P_2O_5/ha/30 cm

The fertilizers were: superphosphate alone, rock phosphate alone and a mixture of these at a ratio of 0.33 to 0.67. Planting in 1976: Merlot/SC.

4. The different ability of the vine varieties to utilize the various nutrients (K, Mg and P) was investigated in field experiments on a sandy and loamy soil with nearly equal nutrient contents.

Results and discussion

1. Pot experiments, Eger (Hungary) 1972–1975

It can be gathered from Table 2 that high P fertilizer amounts with K addition increased the AL–P_2O_5–as well as the EUF–P contents. An increase in the EUF–Mg and EUF–Mn values was also observed, although these nutrients had not been added to the soil. The increase in the EUF–Ca values may be explained by the supply of Ca-containing P fertilizers. Contrary to the EUF–Mg values, the Mg values determined according to Schachtschabel (extraction with $CaCl_2$) do not manifest changes due to P fertilization.

The simultaneous application of high K and P amounts further increased the EUF–P contents, so that they rose clearly above the EUF–P values without K fertilization. The EUF–Ca, EUF–Mg, EUF–Fe and EUF–Mn values were also increased by combined K and P fertilization, although these nutrients were not supplied with the K fertilization. This finding can be explained as follows: K and P fertilizer application increases the concentrations of K^+, Ca^{++} and Cl^- ions in the soil solution, and processes of exchange take place in which the added ions displace other ions occupying exchange positions. Thus Mg, Mn, Fe etc. replace the Ca and K ions in the soil solution, while PO_4 is released into the solution in exchange for the Cl ions. The EUF method is sensitive enough to measure these exchange processes, whereas single chemical extractions cannot distinguish between dissolved and adsorbed ions[4]. Table 3 clearly shows that the nutrient

Table 2. Influence of K and P fertilizer application on the soil nutrient contents as determined by different methods of analysis

Treatments kg/ha/30 cm	EUF		AL	EUF	AL	EUF		Sch**	EUF		
	P*		$P_2O_5^*$	K* 0–30 min	K_2O^*	Ca* 0–30 min	Mg* 0–35 min	Mg*	Fe ppm	Mn ppm	Zn ppm
	0–10 min	0–30 min									
P_0 K_0	0.06	0.20	9	8	28	44	7.5	36	–	–	–
P_{240} K_{730}	0.12	0.48	12	10	34	55	10.7	35	3.25	0.13	0.14
P_{2080} K_0	0.73	2.90	34	12	29	122	15.8	35	1.33	0.23	0.11
P_{2080} K_{3800}	0.91	3.63	36	32	57	150	21.8	39	3.70	0.52	0.13

*   mg/100 g soil
** Sch: according to Schachtschabel

contents in the plant change analogous to the increases in the EUF–P, EUF–K, EUF–Ca, EUF–Mg and EUF–Fe values after K and P fertilizer addition.

The correlation between the EUF–P, EUF–K and EUF–Mg contents on the one hand and the P, K and Mg contents in the vine-stock on the other are very close and highly significant. The regression equations are:

$$y = 0.52 + 0.26x; r = 0.81***; n = 36$$

where
y = K content in d.m. of vine-stock at ripening
x = EUF–K mg/100 g/30 min at 20°C

$$y = 0.135 - 0.426x + 0.862x^2; r = 0.85***; n = 18$$

where
y = P content in d.m. of vine-stock at ripening
x = EUF–P mg/ 100 g/10 min at 20°C

A close correlation also exists between the EUF-K values and yield. In this case the regression equation is:

$$y = 91.31 + 248.68 \log x; r = 0.912***; n = 16$$

where
y = yield (g/vine-stock)
x = EUF–log K/mg/100 g/ 30 min at 20°C

This substantial increase in yield due to increasing K fertilizer levels is remarkable as the soil already had an AL–K_2O value of 28 mg (Table 1). The EUF–K value of the soil was, however, only 8 mg, which explains why yields increased upon K fertilization.

The foliar analysis data of the ripening period correlated much more closely with the EUF values (P, K, Mg) than did the data of the flowering period. The optimal EUF values on the basis of the nutrient contents of leaves can therefore be most effectively assessed at ripening. The optimal levels of the different nutrients during the ripening period are:

Table 3. Influence of K and P fertilizer application on the nutrient contents in the vine-stock

Treatments kg/ha/30 cm	Relative values in leaves							
	P	K	Ca	Mg	Fe	Mn	Zn	N
P_0 K_0	100	100	100	100	100	100	100	100
P_{240} K_{730}	92	113	110	108	135	87	97	114
P_{2080} K_0	142	117	124	138	148	108	110	117
P_{2080} K_{3800}	242	137	131	169	144	117	113	135

P   0.16 − 0.23% dry matter
K   1.00 − 1.40% dry matter
Mg 0.30 − 0.40% dry maater

2. Field experiments, Eger (Hungary) 1976

Tabel 4 shows the changes in the EUF–K and EUF–P values as well as in the AL–K_2O and AL–P_2O_5 values, when different levels of K and P fertilizer were applied to the topsoil as well as to the subsoil. The soils of the treatments 0–8 were only fertilized once, the fertilizer being applied to a depth of 31–60 cm. The topsoil of the treatments 10–16 was fertilized again with the indicated amounts of K and P. Thus, very different K and P supply levels were obtained.

Even the unfertilized control was found to have EUF–K values of 7–9 mg/100 g/30 min at 20°C in the topsoil as well as in the subsoil, whereas the EUF–P values were low.

The vineyard reached the stage of full bearing in 1978 with a yield of 9 t/ha. Since 1979 two loading rates have been used: 6 buds/m^2 and 12 buds/m^2.

Before 1978, no correlation was found between the EUF–K values and the K contents of leaves. Even in the unfertilized controls, the leaves were well supplied with potassium. The K content of leaves in all treatments varied from 1.15–1.80%, both at flowering and ripening. A possible explanation for this finding is that the experimental soil already had EUF–K values of 7–9 mg/100 g in the topsoil as well as in the subsoil (cf Table 4, treatments 1, 5, 9, 13). This is in agreement with results of Bálo et al.[1] who obtained high yields in the variety 'Greenveltliner' with EUF–K values of 8 mg/100 g/30 min at 20°C for a soil depth of 60 cm.

After a yield production of 13–17 t/ha in 1979 the aforementioned EUF–K values were not sufficient for grapes. The K content of leaves had been optimal at flowering, but decreased to a very low level (0.8–0.9%) at ripening, and K deficiency symptoms were observed.

Among the different K fertilizer levels, the first level of 1600 kg/ha proved to be the best. In this case the EUF–K values are about 9–12 mg/100 g/30 min at 20°C and about 7–8 mg/100 g/30–35 min at 80°C. This finding is with results obtained by Németh and Wiklicky[5]. According to these authors, the EUF values recommended for sugar beet are 11–15 mg K/100 g/30 min at 20°C, and indeed 11 mg are required for an EUF–K value of 5 mg at 80°C. In the field experiment in Eger, the EUF–K values at 80°C were 7–8 mg/100 g in the treatment with 1600 kg K fertilizer.

The correlations between the EUF–K values and the K contents in leaves are close and significant, these correlations being notably closer in the subsoil than in the topsoil. The closest correlation was found between the EUF–K values at 80°C in the subsoil and the K contents of leaves. The regression equation is as follows:

$$y = 0.82 + 0.70 \log x; r = 0.71***; n = 16$$

where

y = K content of leaves at ripening

x = EUF–K at 80°C in the subsoil

The correlation coefficient being 0.71*** for a yield of 13 t/ha is improved to 0.85*** for a yield of 17 t/ha.

Table 4 furthermore shows the AL–P_2O_5 and EUF–P values. Since 1976 there was a very close correlation between the EUF–P content and the P content of

Table 4. Quantities of P and K withdrawn by means of AL– and EUF extraction from soils in the various treatments of a field experiment

Treat-ments	Fertilizer applied, kg/ha		Soil depth, cm	Quantities extracted, mg/100 g soil				
	P_2O_5	K_2O		AL–P_2O_5	EUF–P 20°C	AL–K_2O	EUF–K	
							20°C	80°C
1	–	–	0–30	8.0	0.31	28.9	7,1	6.0
	–	–	31–60	6.0	0.18	24.1	5,6	5.2
2	–	–	0–30	9.0	0.44	28.9	6,6	6.1
	–	1600	31–60	6.0	0.20	36.5	5,8	5.8
3	–	–	0–30	7.0	0.13	36.2	8,0	6.4
	–	3300	31–60	4.8	0.15	33.7	7,5	6.4
4	–	–	0–30	6.0	0.18	38.6	8,8	7.5
	–	5000	31–60	4.6	0.14	33.7	9,1	7.6
5	–	–	0–30	10.6	0.76	26.5	8,6	5.5
	1000	–	31–60	9.0	0.32	28.9	6,5	5.2
6	–	–	0–30	11.8	0.77	31.3	8,1	6.3
	1000	1600	31–60	12.4	0.83	38.6	10,8	8.4
7	–	–	0–30	16.0	1.17	38.6	11,2	11.0
	1000	3300	31–60	10.8	0.69	31.3	7,4	7.0
8	–	–	0–30	11.6	1.07	40.5	10,3	11.7
	1000	5000	31–60	7.0	0.12	36.2	8,8	6.4
9	–	–	0–30	8.6	0.20	26.5	6,6	6.4
	–	–	31–60	6.0	0.08	24.1	6.0	4.6
10	–	2000	0–30	9.0	0.32	41.0	12,7	11.0
	–	1600	31–60	6.8	0.07	31.1	9,5	8.6
11	–	3700	0–30	7.8	0.07	62.7	16,8	14.0
	–	3300	31–60	5.0	0.07	47.2	14,8	10.7
12	–	5400	0–30	8.8	0.07	81.9	26,0	18.2
	–	5000	31–60	5.2	0.07	53.0	15,3	11.3
13	1700	–	0–30	41.8	5.45	26.5	12,0	7.3
	1000	–	31–60	23.0	2.36	26.5	8,0	4.6
14	1700	2000	0–60	37.6	3.56	43.4	13,3	12.2
	1000	1600	31–60	22.6	1.81	28.9	10,3	7.5
15	1700	3700	0–30	43.2	4.72	48.2	20,3	17.2
	1000	3300	31–60	24.4	2.33	44.8	14,9	9.2
16	1700	5400	0–30	39.2	4.42	77.1	28,9	18.0
	1000	5000	31–60	25.2	2.31	48.2	16,3	10.7

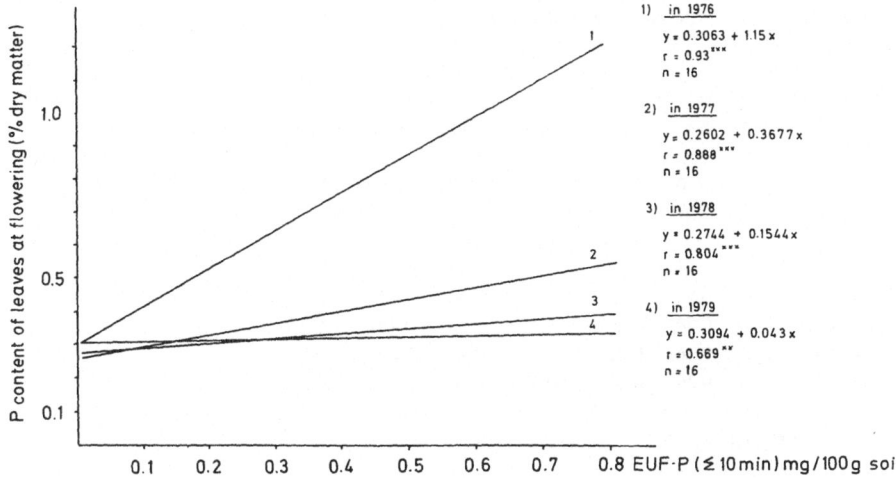

Fig. 1. Correlations between the EUF–P contents desorbed within 0–10 min and the P contents of leaves at flowering in the field experiment.

leaves in the respective years (Fig. 1). The P level of leaves decreased suddenly from 1976 to 1977 and then declined gradually. It may therefore be assumed that the EUF–P values also decline in the course of the years due to ageing of fertilizer phosphates.

3. Pot experiments with different forms of P fertilizer, Eger (Hungary) 1976–1978

Immediately after the application of superphosphate (1976) the EUF–P values increased more than they did after application of hyperphosphate (Table 5). The values for the mixed treatment were intermediate. The P contents of leaves in 1976 showed analogous results (Table 6). No correlation was, however, observed between the $AL-P_2O_5$ values and the P contents of leaves in the year after the fertilizer application.

Each form of P fertilizer influenced the Mn availability in a different way, so that the amounts of Mn taken up by the vinestock were also different (cf Tables 5

Table 5. Influence of different forms of P fertilizer on the size of the EUF–P and EUF–Mn fractions in a pot experiment

Treatment	EUF–P			EUF–Mn
	P_2O_5 /AL/	0–10 min 20°C	0–30 min 20°C	ppm
	mg/100 g soil			
Control	6.2	0.08	0.30	4.0
Superphosphate	42.2	2.25	9.56	5.6
Hyperphosphate	51.6	1.81	6.03	3.2
Mixture	49.2	1.68	7.05	4.0

Tabel 6. Influence of different forms of P fertilizer on the P and Mn contents in the vine-stock in a pot experiment

Treatment	P contents in leaves, % of d.m.					Mn contents in leaves, ppm.				
	1976 Rip*	1977		1978		1976 Rip	1977		1978	
		Flow**	Rip	Flow	Rip		Flow	Rip	Flow	Rip
Control	0.14	0.18	0.17	0.20	0.15	864	227	360	347	395
Superphosphate	0.57	0.25	0.42	0.34	0.49	1279	632	981	477	639
Hyperphosphate	0.26	0.21	0.35	0.40	0.69	188	120	217	172	395
Mixture	0.26	0.21	0.47	0.40	0.58	342	172	332	216	414

* ripening; ** flowering

and 6). The highest Mn uptake was observed in the superphosphate treatment which can be explained by the Ca^{++} ions in the superphosphate having desorbed the Mn^{++} ions. The alkaline hyperphosphate, on the contrary, decreased the Mn availability. In the course of the experimental years, these differences in the Mn contents of vine became smaller.

4. Influence of the variety on the K contents of vine at comparable K levels in the soils

Different white wine varieties (Leányka, O. rizling, Ezerjó) were cultivated in sandy and loamy soils. The K contents were determined in the leaves. The results are shown in Fig. 2.

At nearly equal K levels in the soils, the K contents in the leaves of the variety

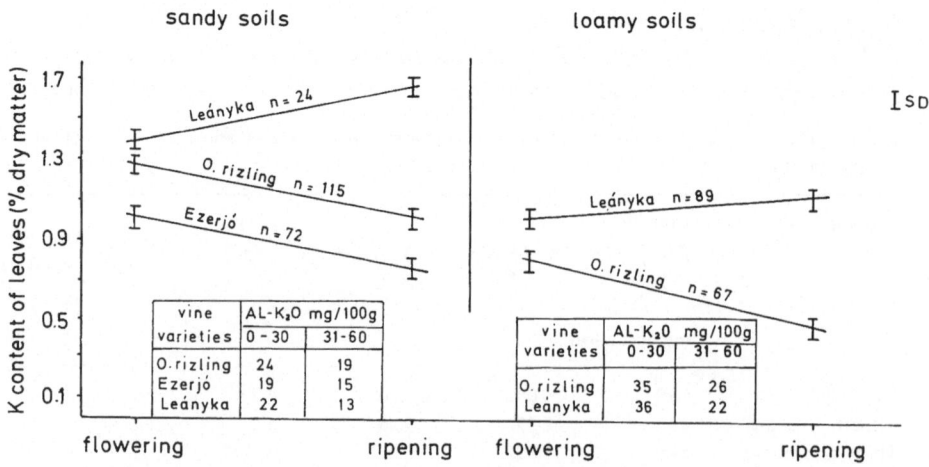

Fig. 2.   K contents of leaves of different white vine varieties in sandy and loamy soils.

'Leányka' increased from flowering to ripening, both on the sandy and loamy soil. The trend was reversed for the rizling variety. This shows that the two varieties have a very different ability to utilize soil K. The same was also observed for Mg, whereas P uptake depended less on variety. The differences in ability of vine varieties to utilixe potassium and magnesium should be taken into account when minimum values for optimal K and Mg nutrition are assessed.

References

1   Bálo E, Pánzel M, Prileszky Gy and Köhalmi I 1980 Effect of heavy amounts of K fertilizer on the K saturation of loess soil, on the N/K ratio of grape leaves and on the quantity and quality of grape yield. Proc. Int. EUF Symp. II, pp 328–336, Budapest, Hungary.

2   Eifert J, Füri J, Szöke L and Várnai M 1976 Praktische Ergebnisse und wissenschaftliche Probleme bei der modernen Nährstoffversorgung von Rebanlagen. Landwirtsch. Forsch. 29, 101–108.

3   Gärtel W 1966 Über die Düngung der Reben in intensiv bewirtschafteten Weinbaugebieten. Weinberg Keller 7, 295–326.

4   Németh K 1976 Die effektive und potentielle Nährstoffverfügbarkeit im Boden und ihre Bestimmung mit Elektro-Ultrafiltration (EUF), Habilitationsschrift Univ. Gießen.

5   Németh K und Wiklicky L 1980 Erfahrungen mit der EUF Methode bei der Düngeberatung. Kali-Briefe Büntehof 15.

Application of the EUF procedure in sugar beet cultivation

L. WIKLICKY

Tulln Sugar Factory, Tulln, Austria

Key words Fertilizer requirement N supply rate Optimal EUF values Sugar beet quality Sugar yield Soil analysis

Summary The relationship between the EUF-nutrient fractions in the soil on the one hand and the nutrient uptake of sugar beet as well as root yield and quality (polarization, α-amino N etc.) on the other is described on the basis of results obtained over several years in surveys conducted in farmers' fields (5000–6000 fields under sugar beet per year) and in field experiments (25–35 sites per year).

Statistically significant close correlations with the respective parameters were found for the following EUF nutrient fractions: $EUF-NO_3$, EUF–P, EUF–K, EUF–Na, EUF–B and EUF–Mn.

Within five years it was possible to determine the EUF-nutrient values which are required for the production of 9 t sugar/ha. These EUF values are the following:

Ca: 65– 70 mg/100 g at 20°C

K: 11– 15 mg/100 g at 20°C (depending on the clay content)

Mg: 3– 5 mg/100 g at 20°C

Na: 2– 3 mg/100 g at 20°C

P: 1.4–1.6 mg/100 g at 20°C

For calculation of the N fertilizer requirements of sugar beet it is suggested to use the sum of the EUF-extractable N amounts. It was found in Austria, Yugoslavia and Denmark over a period of 3 years that the EUF–N value of 1 mg/100 g soil determined between June and September was equivalent to 40 kg N/ha. If, for example, the analysed soil contains 3 mg EUF–N/100 g, $3 \times 40 = 120$ kg N/ha will be available to the sugar beet crop in the following year.

Introduction

The cultivation of sugar beet, the most suitable alternative source of sugar under temperate climate to sugar cane, is determined to a large extent by the interests of the grower on the one hand and by the demands of the processing industry on the other.

Producer of sugar beet

The objective of the farmer is to obtain a maximum of sugar yield per hectare with a minimum of expenditure. In countries where the technological quality (white sugar content, juice purity) is also profitable, the quality aspect plays an important role in the economic calculations of the primary producer.

Sugar industry

The demands of the processing industry are a good sugar content in the root at high technological quality. Moreover, the alkalinity (relation of Σ K + Na to α-amino nitrogen) should be high enough to avoid decomposition of sugars or corrosion in the evaporator by decrease in pH during the processing procedure.

The needs and demands of the sugar beet producer and the sugar industry are characterized by the following parameters:

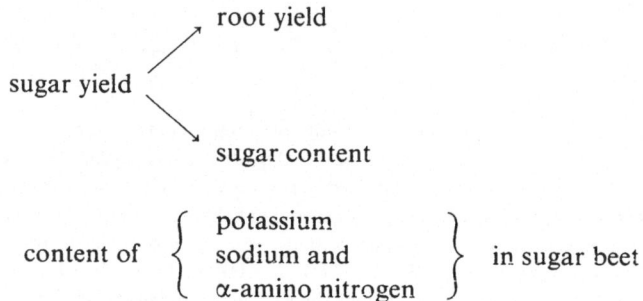

whereby all parameters with the exception of root yield are decisive for the quality of the root.

On the basis of results obtained in the Tulln Sugar Factory it will be attempted to show the influence of those nutrient fractions which are directly or indirectly related to the above parameters. In the Tulln Sugar Factory these parameters are determined as:

$$Q = 99.36 - 0.1427 \text{ (meq K/100 polarization + meq Na/100 polarization + meq } \alpha\text{-N/100 polarization)}$$

(calculated according to the formula of Prof. Wieninger, Austrian Sugar Research Institute, Fuchsenbigl).

Sugar yield (t/ha) = polarization % at the chopping machine × root yield (t/ha)

This sugar yield is not the harvested yield, but the yield which actually remains for processing after deduction of the losses due to storage.

Polarization per cent = sugar content at the chopping machine.

Sugar yield and quality in Lower Austria in the years 1965–1979
Polarization   In Tulln the sugar content at the chopping machine was only about 16.5% between 1965 and 1968 (Fig. 1). A period of optimization followed in the seventies as expressed in the continuous increase in polarization with the exception of an attenuation approximately in the mid-seventies. Even in this latter period the average values were approximately 1.3% of absolute polarization above the lowest value at the beginning of the reference period. In the peak periods, 3-year average values of approximately 19% were reached (moving average values of 3 years were applied to avoid influences of annual weather conditions as far as possible).

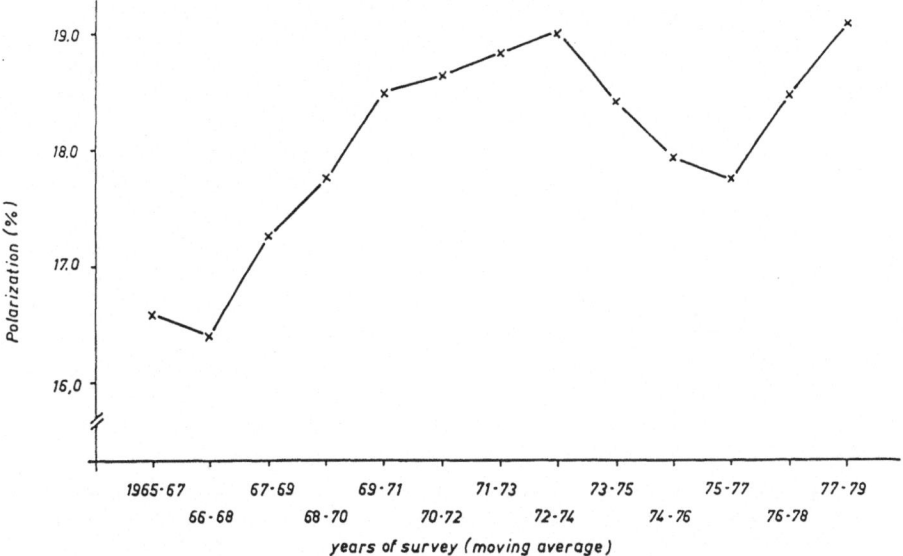

Fig. 1.   Sugar contents in the root of sugar beet in the greater Tulln area from 1965 to 1969.

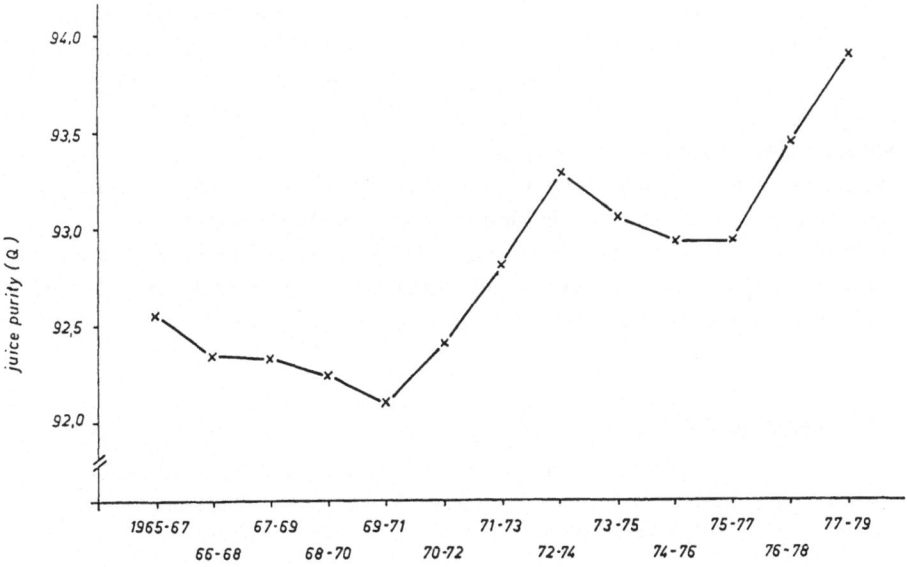

Fig. 2.   Juice purity in the root of sugar beet in the greater Tulln area from 1965 to 1979.

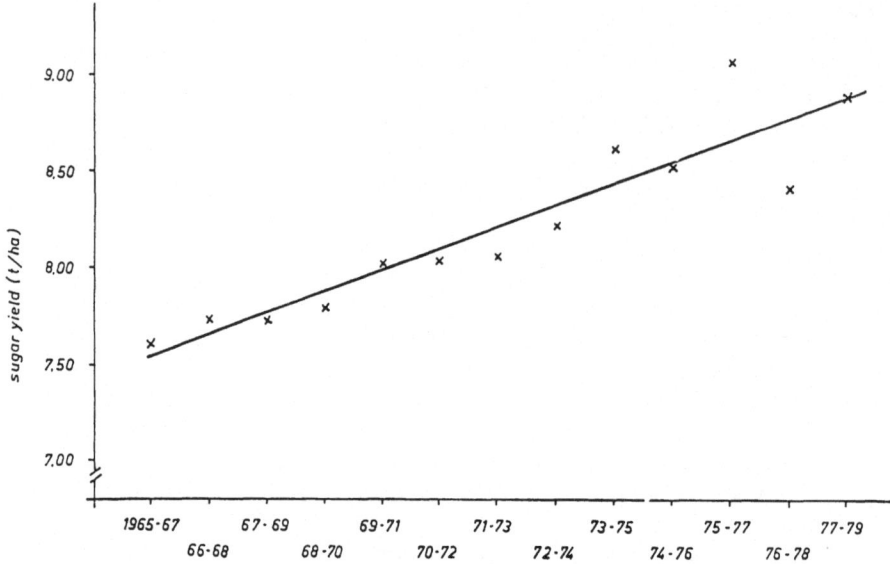

Fig. 3. Sugar yield in the greater Tulln area from 1965 to 1979.

Juice purity (Q) The trend of juice purity is very similar to that of sugar content (Fig. 2). This technologically very important parameter also shows a steady increase and at the end of the reference period reaches with approximately 94% an absolute maximum which will not likely be exceeded.

Sugar yield   It is interesting to note that the improved root quality is linked with the increase in sugar yield (Fig. 3). A permanent increase was observed throughout this period with the exception of the year 1978 in which due to a long dry period the sugar yield amounted to only 7.6 t/ha. The average sugar yield values increased from about 7.6 t/ha in the beginning to about 9 t/ha in the end of the reference period. The sugar yields correspond to the quantities which were delivered to the factory after deduction of loss due to storage.

These results show that Tulln was highly successful in its endeavours to improve the parameters of sugar yield and technological quality. Here, the EUF procedure played an important role as described in the following.

Material and methods

Field experiments and surveys conducted in farmers' fields were evaluated over several years to study the relation between soil data and yields and to assess fertilizer recommendations accordingly. The investigations were carried out by means of electro-ultrafiltration (EUF).

Field experiments

Between 1974 and 1975 numerous field experiments were carried out with sugar beet, wheat, barley and maize to assess the N, P, K, Mg and B fertilizer requirements. From 1978 to 1979 the assessment of nitrogen supply of the soil from the reserves and the effective utilization of mineral fertilizer nitrogen were more thoroughly investigated in field experiments. Furthermore, the application of sugar beet lime sludge, a by-product of sugar beet processing, was subjected to evaluation in the liming practice.

During the whole period of experimentation between 1974 and 1979 32 experiments with cereals (wheat, barley), 5 experiments with maize and 295 experiments with sugar beet have been investigated. All experiments were conducted in five replications according to the randomized block design (evaluation by analysis of variance). In all experiments the harvest was evaluated with regard to yield (cereal grains, roots of sugar beet) and nutrient uptake. The latter was also determined in other plant parts such as straw in cereals and leaves in sugar beet, but not in roots of cereals and maize.

The nutrient contents of the soil were analyzed by means of EUF. The correlations between the EUF nutrient contents on the one hand and the nutrient uptake as well as the yield on the other were statistically calculated.

Surveys conducted in farmers' fields

The values of soil analysis (EUF method) in 1974 and 1975 were compared with the yields and quality values in the respective following years (1975–1976) for 5000 to 6000 fields under sugar beet. Fields with a comparable yield potential were combined to larger units to achieve a greater uniformity as a safeguard against variations due to unknown factors. The results were evaluated according to programs introduced by Prof. Dr. Geidel, Stuttgart-Hohenheim.

Results

Field experiments

EUF-nitrogen fractions   Optimal N supply of sugar beet during the vegetation period has a decisive influence on sugar production and root quality, because a close correlation exists between soluble N content in the root (mg N/100 g) and juice purity (Q).

The following correlation was found in 1975:

$$y = 96.19 - 0.108x; r = 0.82***; n = 16$$

where
y = juice purity (Q)
x = soluble N content in the root (mg N/100 g)

In 1975, shortly after the EUF procedure was introduced in Tulln, a statistically significant correlation was found between the $EUF-NO_3$ contents and the soluble N content in the root:

$$y = 21.65 + 3.56x; r = 0.77***; n = 16$$

where
y = soluble N content in the root
x = $EUF-NO_3$ values in the top- and subsoil in spring 1975.

Even more important is the finding that the $EUF-NO_3$ values (soil samples drawn in autumn 1975) are in close correlation with the EUF–NO_3 values in spring 1976:

$$y = 1.86 + 1.120x; \quad r = 0.88***; \quad n = 33 \text{ locations}$$

where

y = EUF–NO_3 values in spring

x = EUF–NO_3 values in autumn of the previous year

This is an important prerequisite for the assessment of the N requirements of sugar beet by means of EUF–NO_3 values in summer and autumn.

In the following years, however, it became evident that the N requirements could be assessed more precisely by using the sum of the EUF-extractable N fractions instead of the EUF–NO_3 values alone[5].

A correlation, though less close, was observed between the N uptake and the soluble-N content in the root:

$$y = 5.78 + 0.125x; \quad r = 0.55; \quad n = 40$$

where

y = N uptake (kg/ha)

x = soluble-N content in the root

This means that not only the quantity of nitrogen absorbed has a decisive influence on the soluble-N content in the root of sugar beet, but also the yield potential of the site. Accordingly, depending on the yield level, a certain quantity of nitrogen taken up may give rise to different concentrations of α-amino nitrogen in the root.

EUF-potassium fractions   The correlations between the EUF–K values at 20°C and the K contents in the root are significant ($r = 0.63**$). If the K reserves at 80°C are taken into account, the coefficient of correlation rises from 0.63** to 0.73***.

The K removal by sugar beet is decisively influenced by the amount of nitrogen absorbed. Hence the correlation between EUF–K values at 20°C + 80°C and K removal (without taking the N removal into account) is highly significant, but only at a correlation coefficient of 0.67***. If the soils are classified into two groups, namely a) soils with high N removal and b) soils with low N removal, the correlation coefficient of EUF–K and K removal for the soils with high N removal is $r = 0.99***$.

The K content of the root increases with increasing N uptake at constant EUF–K value. The same applies to Na.

Highly significant correlations were also observed between K contents in the root (meq K/root) and EUF–K values. The correlation is closer for EUF–K at 20°C (intensity factor) ($r = 0.74***$) than for EUF–K at 20°C + 80°C ($r = 0.69***$).

EUF-sodium fractions   A significant correlation exists between the EUF–Na values and the quantities of Na removed ($r = 0.58*$). If the sites are divided into soils with high and low N removal, the correlation coefficient is improved from 0.58* to 0.84**.

More important for sugar beet cultivation, however, is the finding that polarization decreases with increasing Na contents in the root.

EUF-phosphorus fractions   The amount of P taken up by sugar beet is much lower than the amount of K. Relatively low amounts of available P (1.5 mg/100 g/30 min at 20°C) are therefore already sufficient to meet the P requirements of the crops.

Already in 1975, the EUF–P values of the experimental sites were above 1.0 mg/100 g. Only three sites have lower P values, namely 0.65, 0.81 and 0.95 mg. The correlations between the EUF–P values and the quantities of P removed were, however, still significant. In 1977 all P values were above 1.0 mg/100 g/30 min at 20°C, so that no correlation was found between EUF–P content and P removal.

EUF-manganese fractions   Significant correlations were observed between the EUF–Mn values and the Mn contents in the root or in the leaf of sugar beet as demonstrated in Fig. 4. This confirms the findings of previous investigations carried out with red clover[2]. The finding which is important for sugar beet is that a significant negative correlation exists between polarization and the EUF–Mn values (Fig. 5).

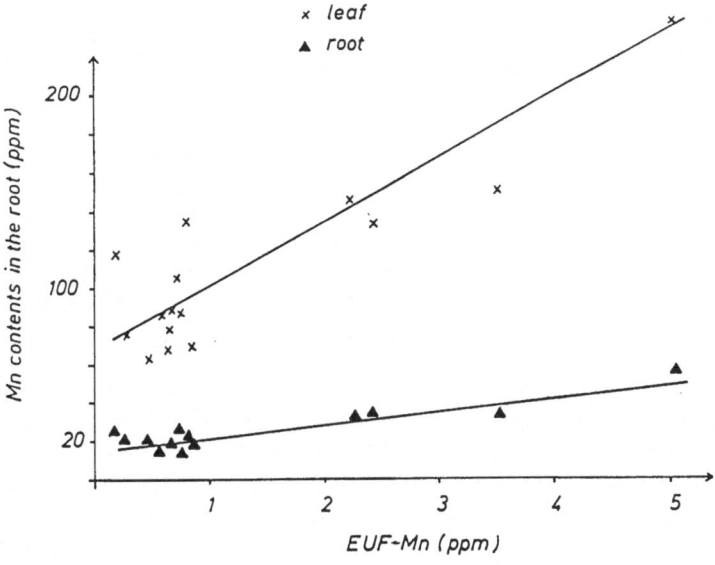

Fig. 4.   Relationship between EUF–Mn values and Mn contents in the roots of sugar beet.

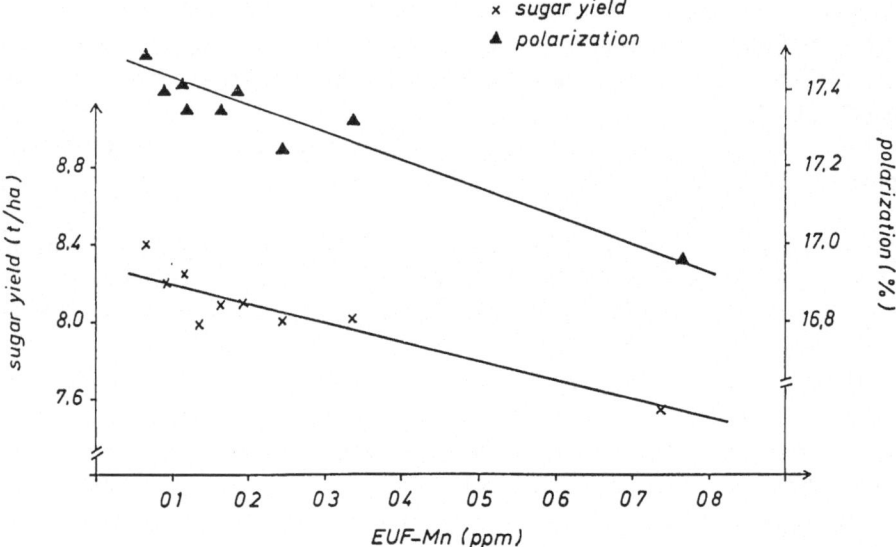

Fig. 5.   Relationship between EUF–Mn and sugar yield as well as the polarization at low K supply.

The negative influence of Mn on polarization can be observed in particular when the K content in the root is low. These results agree with those of previous investigations by Németh and Grimme[1] who found that at constant Mn concentration in the soil solution the negative influence of Mn on the yield of ryegrass increases with decreasing K concentration in the soil solution.

EUF–Mn and EUF–Ca   The negative correlation between the EUF–Mn values and the EUF–Ca values (1st fraction) is highly significant:

$y = 74.41 - 4.8x; r = -0.77***$

$y = $ EUF–Ca (mg/100 g at 20°C)
$x = $ EUF–Mn (ppm)

It is possible to decrease the EUF–Mn contents by liming. The application of carbonation mud was investigated in 3-year experiments. It was found that per ton of carbonation mud (70% DM) the EUF–Ca values in the soil of the 1st fraction (20°C) rose as an average of all test sites by about 2 mg/100 g soil and the EUF–Mn contents decreased continuously.

EUF-boron fractions   A close correlation exists between the EUF–B values and the boron contents of the leaves of sugar beet. This applies to the 1st fraction at 20°C ($r = 0.77***$) as well as to the sum of the 1st and 2nd fraction at 20°C and 80°C ($r = 0.75**$) (Fig. 6.)

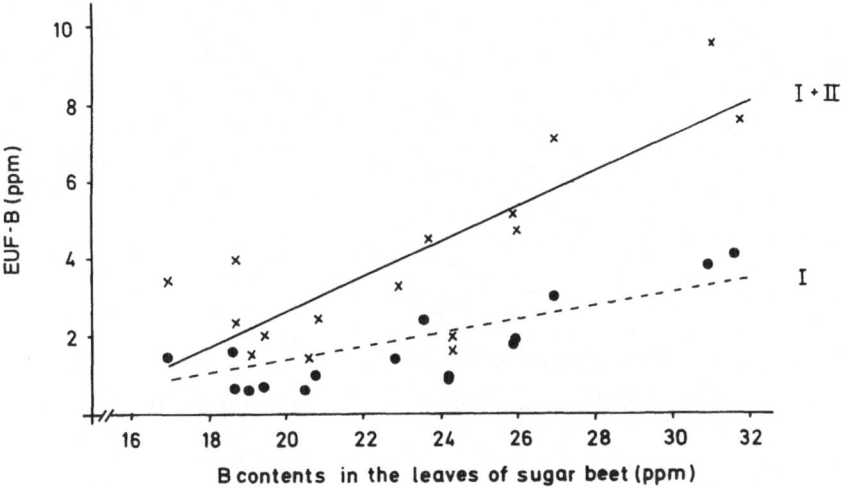

Fig. 6.   Relationship between EUF–B fractions (I and I + II) and the boron contents in the leaves of sugar beet (1979, 16 sites).

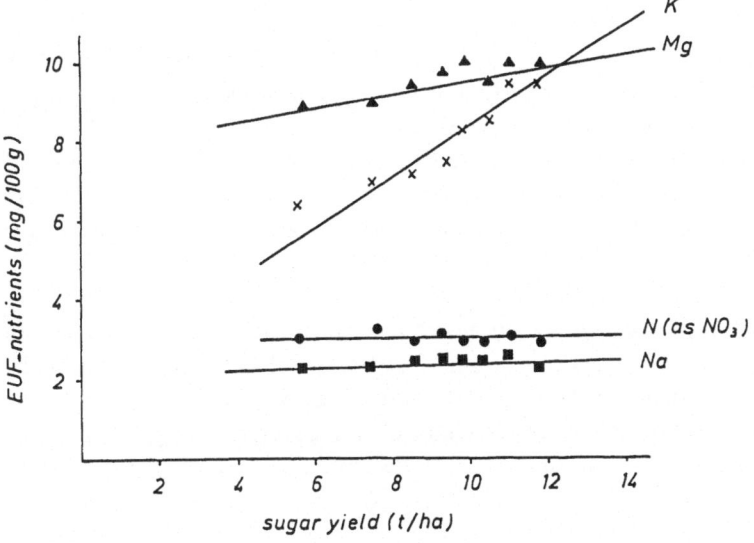

Fig. 7.   Relationship between EUF-nutrient fractions (1974) and sugar yields (1975) from surveys in farmers' fields.

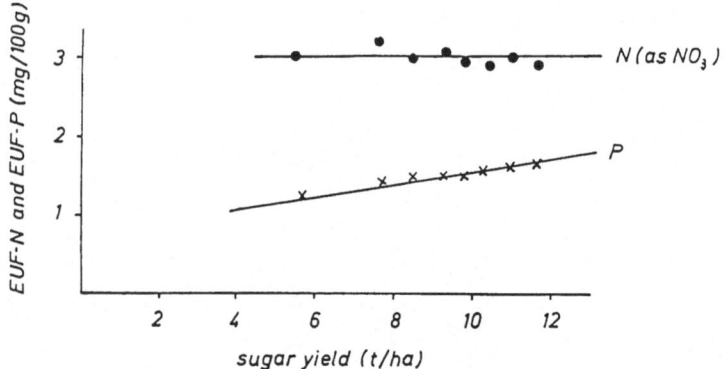

Fig. 8.   Relationship between EUF–N and EUF–P fractions (1974) and sugar yields (1975) from surveys in farmers' fields.

Surveys conducted in farmers' fields
EUF-nutrient fractions and sugar yields (1974 and 1975)   Fig. 7 illustrates the relationship between the EUF–N values which were measured as NO_3 and the EUF–K, EUF–Mg and EUF–Na values in 1974 as well as the sugar yields obtained in the following year on $8 \times 700 = 5600$ fields under sugar beet. The EUF–K and EUF–Mg values have a significant influence on yield, whereas the N and Na values do not affect sugar yields.

A significant influence on yield was also observed for the EUF–P values (Fig. 8). It was therefore attempted to optimize the EUF–K and EUF–P values by adapting the fertilizer recommendations accordingly.

A close positive correlation exists between the EUF–Ca values and sugar yield. A possible explanation is the finding that the EUF–Mn contents decrease with increasing EUF–Ca values. High Mn values have a negative influence on sugar yield as already observed in field experiments.

No significant correlations were found between the EUF–Cu, EUF–Fe and EUF–Zn values on the one hand and sugar yield on the other. This is probably due to the fact that sugar beet is often cultivated on soils where Cu, Fe and Zn deficiency is rarely observed.

EUF-nutrient fractions and polarization   A close correlation exists between the EUF–K and EUF–P values on the one hand and polarization on the other. Polarization rises linearly up to 10 mg K and 1.5 mg P/100 g at 20°C. A linear increase is also observed at increasing EUF–Ca values (up to 60 mg/100 g/20°C). No correlation was, however, found between the EUF–Mg values and polarization.

EUF-nutrient fractions and sugar yields (1976 and 1977)   For the surveys conducted in 1976 and 1977 similar correlations were found between the EUF–

nutrient fractions and sugar yields as already described and demonstrated in Fig. 7 and 8. These figures show that the course of the regression lines changes over the years of experimentation. A possible explanation is that the EUF–K and EUF–P values were increased to the required levels in the course of the vegetation period, so that the rise of the regression lines in Fig. 7 and 8 slowed down accordingly.

Assessment of fertilizer recommendations by means of EUF   When fertilizer requirements are calculated with the use of EUF-nutrient fractions, the clay content and root penetration capability of the site have to be taken into account together with the nutrients released from organic manures as far as they can be determined. The fertilizer recommendations for the 12000–13000 fields under sugar beet were worked out by electronic data processing on the basis of the aforementioned parameters.

EUF threshold values for the production of 9 t sugar/ha
   Ca: 65– 70 mg/100 g soil
   K: 11– 15 mg/100 g soil (depending on the clay content)
   Mg: 3– 5 mg/100 g soil
   Na: 2– 3 mg/100 g soil
   P: 1.5–2.0 mg/100 g soil
   Mn: 0.1–0.3 ppm
   Zn: 0.1–0.3 ppm

Calculation of fertilizer requirements   The K and P requirements are calculated by taking the clay content into account[2,4]. For the calculation of the CaO requirements the results of the experiments with sugar beet lime sludge were used. According to these results, 1 t sugar beet lime sludge (70% DM) increases the EUF–Ca value at 20°C by about 2 mg. It is highly recommended to take the clay content into account as well. The Mg requirements are calculated under the assumption that 30 kg Mg applied per ha increase the EUF–Mg value by 1 mg/100 g.

The boron requirements are calculated with the use of the EUF–B values at 20°C as indicated below:

EUF–B content (ppm)	B requirements (in kg/ha)
0.0 –0.47	4.0
0.48–0.93	3.0
0.94–1.2	2.0

The requirements are calculated by means of the sums of EUF-extractable N which besides NO_3 and NH_4 contain other soluble organic N compounds[5]. Any organic materials applied to the soil after sampling (organic manure, green manuring *etc.*) have to be taken into account as well.

Example: The expected N uptake of sugar beet is approx. 220–250 kg N/ha. One mg EUF–N_{total} corresponds to about 40 kg N/ha. If the soil contains 3 mg EUF–N_{total}/100 g, this means that $3 \times 4 = 120$ kg N/ha are available for uptake. The amount of N fertilizer to be applied to sugar beet is consequently $220 - 120 = 100$ kg N/ha provided no organic N was applied. The effective utilization of fertilizer nitrogen (mineral nitrogen) is assessed at 70% for dressings up to 80 kg N/ha and at 60% for dressings above 80 kg N/ha. This calculation applies to soils with a depth of about 60–100 cm. In soils with a root penetration of less than 60 cm, an addition of about 30–50 kg N/ha is recommended.

This principle of calculation was tested in experiments with increasing N fertilizer levels in 1978 (38 locations) and 1979 (51 locations). Experiments with increasing N nutrition on 16 locations in Denmark of the year 1979 were also evaluated.

Precision of EUF–N fertilizer recommendations in 1978   (38 locations):

21 locations = 55.3%:   Recommendation corresponded to optimal value

10 locations = 26.3%:   Recommendation corresponded to optimal value optimal value, *i.e.* the sugar beet uptake was up to 30 kg N/ha above the optimal value

6 locations = 18.4%:   Recommendation was 51–70 kg N/ha above the optimal value

Precipitation in 1978:   394 mm (extremely dry). These conditions are relatively unfavourable for a reliability test

Root yield: 50 t/ha

Precision of EUF–N fertilizer recommendations in 1979   (51 locations):

36 locations = 70.6%:   Recommendation corresponded to optimal value

9 locations = 17.6%:   Recommendation was 5–50 kg N/ha higher than the optimal value

6 locations = 11.8%:   Recommendation was 51–70 kg N/ha higher than the optimal value

Precipitation: 704 mm

Root yield: 58 t/ha

The higher precision in 1979 is probably due to a better moisture supply (better utilization of native N), taking into consideration that the N release of the soil from the reserves is also characterized by means of EUF.

In the experiments carried out in Denmark the recommended N fertilizer amounts corresponded in 75% of the cases to the optimal N amounts in the roots.

References

1 Németh K and Grimme H 1974 Effect of soil pH on the K concentration in the saturation extract and on the dry matter yield of ryegrass (Lolium perenne). Trans. 10th Int. Congr. Soil Sci. IV, 376–382.

2 Németh K 1976 Die effektive und potentielle Nährstoffverfügbarkeit im Boden und ihre Bestimmung mit Elektro-Ultrafiltration (EUF). Habilitationsschrift, Universität Gießen.

3 Németh K, Makhdum J Q, Koch K and Beringer H 1979 Determination of categories of soil nitrogen by electro-ultrafiltration (EUF). Plant and Soil 53, 445–453.

4 Németh K and Wiklicky L 1980 Erfahrungen mit der EUF-Methode bei der Düngeberatung. Kali-Briefe Büntehof 15.

5 Németh K 1982 Electro-ultrafiltration of aqueous soil suspension with simultaneously varying temperature and voltage. Plant and Soil 64, 7–23.

Factors of plant nutrient availability relevant to soil testing

K. MENGEL

Institute of Plant Nutrition Justus Liebig-University Giessen

Key words   Ammonium Buffering power Nitrate Nutrient availability Phosphate Potassium

Summary   In most arable soils the nitrate availability depends mainly on the quantity of nitrate present in the rooting zone at the beginning of the growing season. Easily mineralizable organic N and the release of non-exchangeable NH_4 from clay minerals may in addition control the nitrogen availability during a season.

In flooded soils, ammonium is the major form of nitrogen absorbed by plants. Ammonium dynamics in these soils is similar to that of potassium. The availability of both is controlled mainly by the intensity and buffering power for ammonium or potassium, respectively.

Basically, intensity of the supply and buffering power for phosphate are the main factors determining the phosphate availability. The determination of the phosphate buffer power, especially in the root zone, however, remains to be difficult.

Soil test methods should take into consideration the major factors and processes relevant to the availability of a particular plant nutrient.

Introduction

Under practical conditions, modern plant nutrition means providing plant nutrients to crops so that optimum crop production is achieved. Such a nutrient supply includes the application of fertilizer. The rates applied should meet the demand of the crop, but should not exceed the demand to any major extent. Hence, fertilizer recommendations should take into account the available nutrients already present in the soil. It is for this reason that soil tests are carried out. A reliable soil testing method should provide precise information on the fertilizer quantity required for optimum growth of a particular crop.

Presently, the use made of soil testing is still far from complete. One of the reasons for this unsatisfactory situation is that nutrient availability in a soil depends on many factors and processes. This is true for the most important plant nutrients: nitrate, ammonium, potassium and phosphate. In the following, processes and factors relevant to the availability of these nutrients are considered in the light of the dependence of modern plant nutrition on soil testing.

Nutrients

Nitrate

It is now generally accepted that the concept of Barber et al.[4] is a sound approach in considering nutrient availability. These authors stated that only a very small portion of each of the major nutrients (N, P, K) is accessible to plant roots by interception, but that by far the greatest portion of these nutrients reaches the plant roots by mass flow and diffusion. Mass flow is of major importance for nutrients which occur at relatively high levels in the soil solution[2] so that the nutrient concentration at the root surface is only slightly suppressed by the nutrient uptake of the plant. This is generally the case for nitrate, which is hardly bound to soil particles and easily soluble in the soil solution.

In contrast to phosphate and potassium, of which the movement through soil can be affected by precipitation and adsorption processes, the mobility of nitrate is hardly influenced by such processes. The most important factor influencing nitrate mobility is soil moisture. Under low soil moisture conditions, the water flux towards the roots is impaired and hence also the rate of nitrate transport is affected. The relationship between nitrate availability in the soil and soil water content has been clearly shown by model experiments of Casper[8]. Fig. 1 shows this relationship for 4 nitrate application rates (0, 40, 80, 120 mg nitrate-N/kg soil). Optimum N uptake of maize plants was obtained at a water content of 19 percent (= pF 2.0). The finding of this experiment is consistent with observations under field conditions, showing that in dry periods a nitrate accumulation in the surface soil layer may occur[29,35]. A further factor which is important for the utilization of soil nitrate is the rooting pattern and rooting density.

Under normal growth conditions (no drought, normal root growth) crops are capable of depleting the nitrate in the rooting zone to a very low level. Thus according to numerous field experiments of Wehrmann and Scharpf[34], the

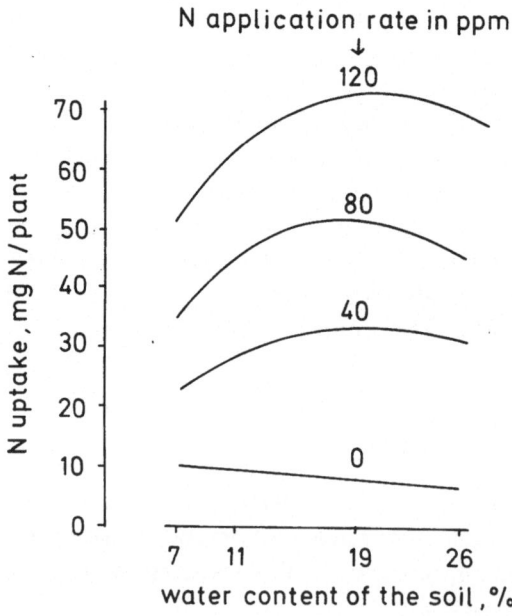

Fig. 1.   Effect of soil moisture on the uptake of nitrate by maize (N quantity in maize shoots).

nitrate quantity generally found in June under winter wheat amounted to not more than about 20 kg N/ha. Also sugar beets are remarkably capable of depleting the nitrate in the soil[36]. According to these results, about 80 to 90 percent of the nitrate present in the rooting zone is accessible to plant roots. It is for this reason that nitrate availability is primarily a question of nitrate quantity and less a question of nitrate intensity, *e.g.* nitrate concentration in the soil solution. This nitrate quantity in the soil profile (depth of the rooting zone) can be assessed at the beginning of the growth period and N fertilizer rates can be recommended on the basis of such soil tests.

This procedure has been proposed by various authors[25,33] and has been well developed in West Germany by Wehrmann and Scharpf[34].

The procedure, however, only takes into account the nitrate quantity which is already present in the soil at the beginning of the growth period and does not consider the nitrate quantities which will be produced during the growth period by microbial activity. In loess soils, the nitrate thus produced during the season is considered to be in the order of 20 kg N/ha[5], but according to Winner *et al.*[36] may also amount to about 50 kg N/ha. The quantity is particularly important in soils with higher amounts of organic N and thus with a high nitrate production potential. In such soils, the assessment of nitrate in the soil profile at the beginning of the growth period may lead to wrong N fertilizer recommendations[14]. It is for this reason that a reliable soil test for available nitrogen should be based not only on the nitrate quantity present at the beginning of the growth season, but should also give some information on the potential of easily mineralizable organic N.

According to recent results of Scherer and Mengel[32] some soil types may contain considerable amounts of NH_4–N fixed by 2:1 clay minerals. This so-called fixed NH_4 was in the range of 1500 to 3000 kg N/ha. Still unpublished experimental results of these authors provide evidence that some of this non-exchangeable NH_4 is available to crops. On such soils, soil tests should also take into account the non-exchangeable NH_4.

Ammonium

In most well aerated soils, NH_4–N is rather quickly oxidized to NO_3^- so that NH_4–N plays only a minor role as direct plant nutrient in these soils. Thus, generally, the quantities of unspecifically bound NH_4 are low (< 10 kg N/ha) in arable soils[32,34]. This is the reason why for the assessment of the available nitrogen in such soils a determination of nitrate is often sufficient.

In flooded soils, however, the situation is quite different. Here, any nitrate present during flooding is quickly denitrified[31] and NH_4–N is the main nitrogen source upon which the crop is feeding. Thus, for wetland rice soils the available NH_4 and also the easily mineralizable organic nitrogen in the soil are important components of nitrogen availability. Routine analysis carried out in wetland rice soils are based on the determination of nitrogen quantities rather than on intensities[9]. The behaviour of NH_4 in flooded soils is related more to that of K than to that of nitrate. Potassium availability is more a question of K intensity and less a question of quantity. It is therefore suggested that in assessing the available NH_4–N in wetland rice soils also the intensity factor should be taken into consideration. This suggestion is well supported by the results of Wanasuria presented in this symposium, who found that the EUF-extractable soil NH_4 rather than the exchangeable NH_4 controlled the N uptake and grain yield of the rice crop. All considerations presented below for K are in principle also applicable to NH_4 in flooded rice soils.

Potassium

The so-called available K in soil, e.g. the K extracted with the 'Doppellaktatmethode' or by an exchange with NH_4, generally exceeds by far the quantity required by the crop. Thus according to field experiments of Loué[20] on mediumtextured soils, a winter wheat crop requires at least a level of 200 ppm exchangeable K which amounts to about 600 kg K/ha. The K uptake by the crop, however, is only in the range of 120 to 150 kg K/ha. This example may show that for an optimum K supply to the crop, the amount of exchangeable and thus easily available K must be several times higher than the K uptake of the crop. The reason for this finding is that the transfer of K from the various soil regions towards plant roots often can become the limiting proces in the K supply.

It is now generally accepted that the most important process through which K moves towards plant roots is diffusion. The diffusivity of the soil medium, which is strongly influenced by soil moisture[23] is therefore one important factor controlling the movement of K to the plant roots. The other important factors are the K concentration in the bulk soil solution and the K buffering capacity. Potassium uptake by plants can be described by the following equation[11]:

$$U = 2\pi a\alpha \cdot \bar{c}t$$

where
U = uptake
a = Radius of the root
α = Root absorbing power
$\bar{c}$ = Average K concentration at the root surface
t = Duration of absorption period

The equation refers to the K uptake of a 1-cm long root section. In this context, the root radius (a) and the root absorbing power (α) will not be discussed in much detail. It should be emphasized that the K uptake is related to the K concentration at the root surface in a linear way. This is not completely correct, as the K-uptake rate in relation to the K concentration in solution is described more closely by a Michaelis-Menten type of curve, as was shown by Claassen and Barber[10]. In cases, however, in which diffusion is the limiting process in transporting K to the root surface, the K concentration at the root surface is low ($< 50 \mu M$) and in this low concentration range the relationship between K uptake and K concentration is approximately linear.

Determining the K concentration at the troot surface is extremely difficult. It is prone to quick variations due to the K-absorbing power of the root, which strongly depends on root metabolism[21] and also on the K content of plants[3]. The K concentration at the root surface is closely related to the K concentration in the bulk soil solution[1]. This K concentration is a reproducible term, if measured under standard conditions, e.g. constant soil moisture content.

The term c̄ in the abovementioned equation is an 'intensity' term and refers to the average K concentration at the root surface during a certain uptake period. At the beginning, the K concentration at the root surface is higher than at the end of the uptake period. This decline in K concentration at the root surface also depends on the capability of the soil to replenish the soil solution K. This potential of replenishing the solution K is identical with the K-buffering capacity. It thus appears that the above cited equation also includes the K-buffering capacity because the 'c̄ level' is influenced by K replenishment. The K-buffering capacity is described by the following equation[22]:

$$b = \frac{\Delta K \; (Quantity)}{\Delta K \; (Intensity)}$$

where b = buffering capacity

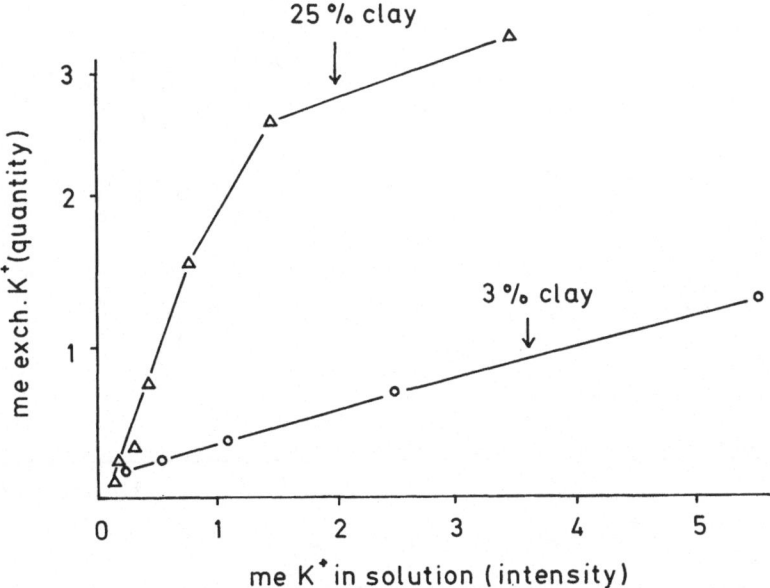

Fig. 2.   K⁺ buffer curve of a sandy soil and a clay soil.

'K-Quantity' generally is measured as exchangeable K, although also the fraction of non-exchangeable soil K may contribute to the replenishment of the solution K [24]. 'K-Intensity' refers to the equilibrated K concentration in the soil solution. The K-buffering capacity of a soil is thus represented by a curve the steepness of which is a measure for the buffer power. The curves of Fig. 2, established from research data of Grimme et al. [13] show the K-buffering capacity of a sandy soil and a clay soil. It appears that the latter has a much higher K-buffering power than the sandy soil. The curve for the clay soil is not linear which indicates that adsorption sites of different K specifity (p-, e-, i-positions) are involved and that the K-buffering power in the low K concentration range (< 2 meq K/l) is higher than in the concentration range > 2 meq K/l.

Although the K-buffering capacity is an important factor in characterizing K availability, it hardly has been used in routine soil testing. Németh[26], in investigating 3 soils with different K-buffering capacities found a close negative relationship between the K-buffering capacity and the decrease in grass yield of 4 consecutive cuts harvested during the experimental period. Von Braunschweig[6] indirectly takes into account the K-buffering capacity by relating the lactate-soluble K to the clay content. The 'K availability data' thus obtained were in good agreement with K-fertilizer response obtained in numerous field experiments. Recent experimental results of During and Duganzich[12] show that the K uptake by white clover was mainly controlled by the K concentration in the soil solution and the K-buffering capacity. This result is in good agreement with experimental data of Busch[7] which provide evidence that the level of the

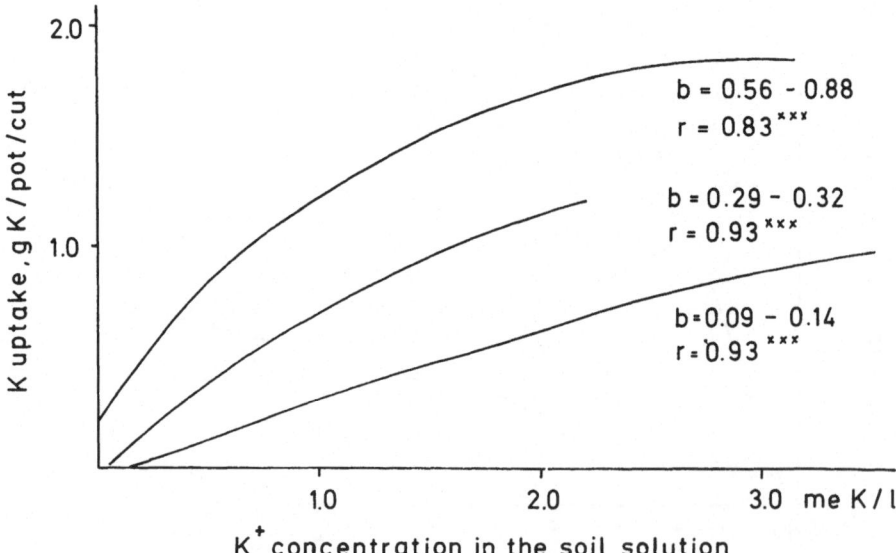

Fig. 3.   Relationship between K $^+$ buffer capacity, K $^+$ concentration of the soil solution, and K $^+$ uptake of *Lolium multiflorum*.

optimum K concentration in the soil solution depends on the K-buffering capacity of the soil. The soils involved in these investigations were grouped into three categories: low, medium and high K-buffering capacity. Fig. 3 depicting the relationship between the K uptake by grass and the K concentration in the soil solution for these 3 soil groups shows that the curve for the high K-buffering capacity levelled off at a low K concentration and *vice versa*.

Phosphate

Analogous to the K availability, that of phosphate is also controlled mainly by the diffusive flux of phosphate from the surrounding soil medium towards the root. All what has been said in the above section about the important availability parameters such as soil moisture, nutrient concentration in the soil solution and buffering capacity is also true for the phosphate availability. The importance of these parameters for phosphate uptake has been shown in model experiments by Olsen and Watanabe[28] as well as by Holford[16]. In one point, however, a substantial difference exists between phosphate and K availabilities. Concerning the process of equilibrating, the equilibrium for K in the soil medium is obtained relatively quickly, at least if the soil is not a K-fixing soil [17], whereas the phosphate equilibration process extends over months or even years[19, 30.] Thus a phosphate buffer curve established now may shift remarkably during the following months towards more phosphate fixation or more phosphate solubilization depending on soil conditions. One of the most important soil factors in this respect is soil pH[15, 18].

Measuring the pH in the bulk soil may be of little help as the zone around the plant root from which phosphate is drawn by the plant can significantly differ in its pH from the pH of the bulk soil. Because of this dilemma, establishment of the actual phosphate-buffering power in the phosphate depletion zone around the root is hard to achieve. This is one of the main reasons why the concept of concentration and buffering capacity so well applicable to the K availability is not always satisfactory in predicting phosphate availability[27]. According to recent, still unpublished results of Keerthisingh, lactate-soluble phosphate and EUF-extractable phosphate were more closely related to the phosphate uptake of wheat than were phosphate concentration in the soil solution and the phosphate-buffering power. It is therefore doubtful whether for routine soil analysis, an assessment of the phosphate concentration in the soil solution and of the phosphate-buffering capacity will be a suitable approach towards estimating phosphate availability. Until now, the determination of the phosphate solubility in the soil by conventional methods or by the EUF technique seems to be the most reliable, but not yet satisfactory, approach towards the estimation of phosphate availability.

Conclusions

The basic processes controlling plant nutrient availability differ for the major nutrients. Nitrogen availability in arable soils is mainly a function of nitrate quantity and easily mineralizable organic N in the rooting zone of the soil. Thus, nitrogen soil testing should focus on these parameters. For NH_4 (flooded soils), K and phosphate, the intensity factors are the most important parameters controlling availability. Hence these parameters should be taken into account in soil testing methods. Assessment of the nutrient availability includes the determination of the nutrient concentration in the bulk soil solution as well as the nutrient-buffering capacity, since mainly these factors control the nutrient flux into roots. The determination of the 'phosphate intensity' still meets with difficulties. Particularly the conditions in the rhizosphere, controlling phosphate solubility, adsorption and desorption, are not yet well understood. More research is still required to elucidate this essential aspect of phosphate availability. Information thus obtained should be the basis for the development of more reliable phosphate soil tests.

References

1   Baldwin J P, Nye P H and Tinker P B 1973 Uptake of solutes by multiple root systems from soil. Plant and Soil 38, 621–635.
2   Barber S A 1974 Influence of the plant root on ion movement in soil. *In* The Plant Root and Its Environment. Ed. E W Carson. University Press of Virginia, Charlottesville 525–564.
3   Barber S A 1979 Growth requirement of nutrients in relation to demand at the root surface. *In* The Soil-Root Interface. Eds. J. L. Harley and R S Russell. Academic Press, London pp 5–20.
4   Barber S A, Walker J M and Vasey E H 1963 Mechanism for the movement of plant nutrients from the soil and fertilizer to the plant root. J. Agric. Food Chem. 11, 204–207.
5   Böhmer M, Scharpf H C and Wehrmann J 1977 Mineralstickstoffvorrat und -nachlieferung im Boden, Komponenten der Stickstoffversorgung der Pflanzen, Landwirtsch. Forsch. Sonderh. 34, Kongreßband 45–54.
6   Braunschweig L C v 1978 Ergebnisse aus mehrjährigen Feldversuchen zur Überprüfung der optimalen Kaliversorgung des Bodens. Landwirthsch. Forsch. Sonderh. 35, 219–231.
7   Busch R 1980 Der Einfluß der K$^+$-Konzentration der Bodenlösung und der K$^+$-Pufferung auf die K$^+$-Aufnahme und das Wachstum von *Lolium multiflorum*. Diss. FB 19, Justus Liebig-Universität Gießen.
8   Casper H 1975 Nitratverfügbarkeit als Funktion des Wassergehaltes im Boden. Ein Modellversuch mit jungen Maispflanzen. Diss. FB 19, Justus Liebig-Universität Gießen.
9   Chang S C 1978 Evaluation of the fertility of rice soils. *In* Soils and Rice. The Intern. Rice Res. Institute, Los Banos pp 521–541.
10   Claassen N and Barber S A 1976 Simulation model for nutrient uptake from soil by a growing plant root system. Agron. J. 68, 961–964.
11   Drew M C, Nye P H and Vaidyanathan L V 1969 The supply of nutrient ions by diffusion to plant roots in soil. I. Absorption of potassium by cylindrical roots of onion and leek. Plant and Soil 30, 252–270.
12   During C and Duganzich D M 1979 Simple empirical intensity and buffering capacity measurements to predict potassium uptake by white clover. Plant and Soil 51, 167–176.

13   Grimme H, Németh K and Braunschweig L C v 1971 Beziehungen zwischen dem Verhalten des Kaliums im Boden und der K-Ernährung der Pflanze. Landwirtsch. Forsch. Sonderh. 26, 165–176.

14   Gutser R 1978 Paper presented at DLG-Wintertagung in Wiesbaden.

15   Hagemann O and Müller S 1976 Untersuchungen über den Einfluß des pH-Wertes auf die Ausnutzung von Düngerphosphaten und die Mobilisierung von Bodenphosphaten. Arch. Acker. Pflanzenbau Bodenkd. 20, 805–815.

16   Holford I C R 1976 Effects of phosphate buffer capacity of soil on the phosphate requirement of plants. Plant and Soil 45, 433–444.

17   Karbachsch M 1978 Kaliumernährung des Tabaks auf einem K-fixierenden nordwestiranischen Boden. Z. Pflanzenernaehr. Bodenkd. 141, 513–522.

18   Keerthisinghe G 1980 Die Bedeutung der Phosphatpufferung und der Phosphatkonzentration der Bodenlösung für die Phosphataufnahme der Pflanze. Eine Untersuchung an Weidelgras und Sommerweizen auf repräsentativen mitteleuropäischen Böden. Diss. FB 19, Justus Liebig-Universität Gießen.

19   Keerthisinghe G and Mengel K 1979 Phosphatpufferung verschiedener Böden und ihre Veränderung infolge Phosphatalterung. Mitt. Deutsch. Bodenkd. Ges. 29, 217–230.

20   Loué A 1979 Die durchschnittliche Wirkung der Kalidüngung auf Großkulturen in Dauerversuchen. Kali-Briefe Bern, Fachgeb. 16, 79. Folge, Nr. 4.

21   Mengel K 1967 Moderne Gesichtspunkte in der Pflanzenernährung und Düngung. Schriftenr. Landwirtsch. Fak. Gießen, Heft IV, 73–82.

22   Mengel K 1974 Die Faktoren der Kaliverfügbarkeit und deren Bedeutung für die Ertragsbildung. Landwirtsch. Forsch. Sonderh. 31, 45–58.

23   Mengel K and Braunschweig L C v 1972 The effect of soil moisture upon the availability of potassium and its influence on the growth of young maize plants (Zea mays L.). Soil Sci. 134, 142–148.

24   Mengel K and Wiechens B 1979 Die Bedeutung der nicht austauschbaren Kaliumfraktion des Bodens für die Ertragsbildung von Weidelgras. Z. Pflanzenernaehr. Bodenkd. 142, 836–847.

25   Müller S, Ansorge H, Hagemann O, Görlitz H, Garz J and Stumpe H 1976 Untersuchungen über die Möglichkeit einer Bemessung der ersten N-Gabe zu Getreide durch Berücksichtigung des Gehaltes an anorganischem Stickstoff im Boden. Arch. Acker Pflanzenbau Bodenkd. 20, 713–722.

26   Németh K 1975 The effect of K fertilization and K removal by rye-grass in pot experiments on the K concentration of the soil solution of various soils. Plant and Soil 42, 97–107.

27   Nye P H 1979 Soil properties controlling the supply of nutrients to the root surface. In The Soil-Root Interface. Eds. J L Harley and R S Russell. Academic Press, London, New York, San Francisco, pp 39–49.

28   Olsen S R and Watanabe F S 1970 Diffusive supply of phosphorus in relation to soil texture variations. Soil Sci. 110, 318–327.

29   Page M B and Talibudeen O 1977 Nitrate concentrations under winter wheat and in fallow soil during summer at Rothamsted. Plant and Soil 47, 527–540.

30   Pagel H and van Huay H 1976 Wichtige Parameter der Phosphat-Sorptionskurven einiger Böden der Tropen und Subtropen und ihre zeitliche Veränderung durch P-Düngung. Arch. Acker Pflanzenbau Bodenkd. 20, 765–778.

31   Ponnamperuma F N 1972 The chemistry of submerged soils. Adv. Agron. 24, 29–96.

32   Scherer H W and Mengel K 1979 Der Gehalt an fixiertem Ammonium-Stickstoff auf einigen repräsentativen hessischen Standorten. Landwirtsch. Forsch. 32, 416–424.

33   Soper R J and Huang P M 1962 The effect of nitrate nitrogen in the soil profile on the response of barley to fertilizer nitrogen. Can. J. Soil Sci. 43, 350–358.

34   Wehrmann J and Scharpf H C 1979 Der Mineralstoffgehalt des Bodens als Maßstab für den Stickstoffdüngerbedarf (N_{min}-Methode). Plant and Soil 52, 109–126.

35 Weller F 1971 Nitrate in Böden unter Intensivkulturen. *In* Hohenheimer Arbeiten, Heft 58, Vorträge des Seminars Umweltforschung der Universität Stuttgart-Hohenheim pp 50–55.

36 Winner C, Feyerabend I and Müller A v 1976 Untersuchungen über den Gehalt an Nitratstickstoff in einem Bodenprofil und dessen Entzug durch Zuckerrüben. Zucker 29, 477–484.